LEARNING FROM AGRI-ENVIRONMENT SCHEMES IN AUSTRALIA

INVESTING IN BIODIVERSITY AND OTHER ECOSYSTEM SERVICES ON FARMS

Part II. The birds and the beef

Part III. Planning, doing and learning

Participants at the 2014 workshop on agri-environment schemes in Canberra.

From left to right: Back row — Angela Newey, Bill Woodruff, Geoff Kay, Graeme Doole, Stuart Whitten, Geoff Park, Sayed Iftekhar, Maksym Polyokov, Dean Ansell, and Emma Burns. Front row — David Salt, Rob Fraser, Graham Fifield, David Duncan, David Pannell, Fiona Gibson, and Phil Gibbons.

Source: Photo by David Salt.

Preface

In times gone by, environmental issues on farmland were seen largely through the lens of agricultural production — we viewed the problems and their possible solutions in terms of how they affected agricultural outputs. Weeds, pests and erosion were challenges because they reduced the land's productivity.

In recent decades, it has been recognised that farms can deliver much more than the sum of their agricultural products. In Australia, farming landscapes cover more than half our land mass, they provide refuge for many unique native animals and plants, and they are home to people.

In acknowledgement of the multiple values associated with farmland, Australian governments have been paying farmers to provide public goods and services for many years. These goods and services include habitat for wildlife, and healthier soil and water quality. Many countries around the world have been doing this too. These government programs have been commonly referred to as agri-environment schemes, as they are about improving environmental values in an agricultural space.

The Australian experience goes back a little over a quarter of a century, with well over $7 billion of public money having been invested. Unfortunately, Australia's National Audit Office has found (repeatedly) that the programs have been unable to demonstrate enduring environmental outcomes. Over the same period, the nation has seen continuing declines in biodiversity, and land and water quality. To turn this around, can we learn from what has been done in the past (both here and overseas)?

Agri-environmental policy is an inherently complicated beast, involving a raft of different players, from farmers and conservationists to taxpayers and politicians. Each group brings with it a diverse set of motivations and interests, including maximising profit, minimising biodiversity loss, and everything in between. A proper evaluation of agri-environment policies, therefore, requires a multi-disciplinary approach.

Towards the end of 2014, a group of people interested in biodiversity conservation and agriculture—ecologists, economists, social scientists, practitioners, and policymakers — met at a workshop in Canberra to share their knowledge and experience of agri-environment schemes in Australia. This book draws together the diverse experiences, ideas, and perspectives presented at that meeting.

Each chapter presents a different perspective on the challenge of designing and running effective agri-environment schemes. For anyone with an interest or a stake in agri-environment investment in Australia or overseas, we are confident there will be many lessons and insights for you in the following pages.

The workshop on lessons from agri-environment schemes was sponsored by the National Environmental Research Program Environmental Decisions Hub with additional financial support from Phil Gibbons (The Australian National University) and Dave Pannell (University of Western Australia). Logistics support was provided by Jane Campbell (University of Queensland). We are also indebted to the Spillers, owners of 'Woodlands', for showing us through their Whole of Paddock Rehabilitation site, and also Mick Woods, manager of 'Woolooware', for taking the time to show us a Box Gum Grassy Woodland Project site of the Environmental Stewardship Program. Lastly, we also thank Graham Fifield and Greening Australia for providing refreshments and guidance when we visited these two field sites.

Dean Ansell, Fiona Gibson, and David Salt[1]

---

1   Dean, Fiona, and David jointly managed the workshop that led to this book and shared the jobs of editing and authoring the introductory and concluding chapters. Each claims an equal share of the book and names are ordered alphabetically.

LEARNING FROM AGRI-ENVIRONMENT SCHEMES IN AUSTRALIA

INVESTING IN BIODIVERSITY AND OTHER ECOSYSTEM SERVICES ON FARMS

Edited by Dean Ansell, Fiona Gibson and David Salt

Australian
National
University

PRESS

ANU PRESS

Published by ANU Press
The Australian National University
Acton ACT 2601, Australia
Email: anupress@anu.edu.au
This title is also available online at press.anu.edu.au

National Library of Australia Cataloguing-in-Publication entry

Title:	Learning from agri-environment schemes in Australia : investing in biodiversity and other ecosystem services on farms / Dean Ansell, Fiona Gibson, David Salt.
ISBN:	9781760460150 (paperback) 9781760460167 (ebook)
Subjects:	Ecosystem services--Australia. Incentives in conservation of natural resources--Australia. Biodiversity--Government policy--Australia Biodiversity--Economic aspects--Australia.
Other Creators/Contributors:	
	Ansell, Dean, editor. Gibson, Fiona, editor. Salt, David, editor.
Dewey Number:	333.95160994

Cover design and layout by ANU Press. Cover photograph by David Salt.

Contents

Part I. The agri-environment in the real world

Part II. The birds and the beef

Part III. Planning, doing and learning

List of figures

List of tables and boxes

List of acronyms and abbreviations

ABARES	Australian Bureau of Agricultural and Resource Economics and Sciences
BCR	Benefit–cost ratio
BMP	Best management practice
CAP	Common Agricultural Policy
CfoC	Caring for our Country
CMA	Catchment management authority
CT	Conservation tender
eNGO	Environmental non-government organisation
EPBC	*Environment Protection and Biodiversity Conservation Act 1999*
EU	European Union
INFFER	Investment Framework for Environmental Resources
LSLS	Land Sharing Land Sparing
MBIs	Market-based instruments
MERI	Monitoring, Evaluation, Reporting, Improvement
NAP	National Action Plan for Salinity and Water Quality
NHT	Natural Heritage Trust
NHT2	Second round of the Natural Heritage Trust
NRM	Natural resource management

OECD	Organisation for Economic Co-operation and Development
SMART	Specific, Measurable, Attainable, Relevant, and Time-Bound
WOPR	Whole of Paddock Rehabilitation
WQIP	Water Quality Improvement Plan
WWF	World Wildlife Fund

Contributors

Dean Ansell

Dean Ansell is a PhD student in the Fenner School of Environment and Society at The Australian National University. His PhD focuses on the cost-effectiveness of biodiversity conservation in agricultural landscapes. His research also involves on-ground evaluation of the cost-effectiveness of ecological restoration projects in farmland in south east Australia. He has more than 15 years of experience working with government and non-government organisations on biodiversity conservation and natural resource management in Australia and internationally.

Simon Attwood

Simon Attwood is an agroecological scientist with Bioversity International. His current work focuses on managing ecosystem services to sustainably intensify production for poverty reduction and food security outcomes. Previously, Simon worked within the Biodiversity Conservation Branch in the Department of the Environment, Water, Heritage and the Arts, helping to design and deploy the Environmental Stewardship Program. Simon has a PhD examining arthropod responses to land-use intensification.

Louise Blackmore

Louise Blackmore is a PhD student in the School of Agricultural and Resource Economics at the University of Western Australia. Her key research interest is the socio-economic aspects of biodiversity conservation. Louise's PhD studies use experimental economics methods to explore collaborative management of biodiversity by private landholders in Australia.

Emma Burns

Emma Burns is Executive Director of the Australian Long Term Ecological Research Network, and a conservation biologist within the Fenner School of Environment and Society, The Australian National University. Since completing her doctoral research on conservation genetics and phylogeography, Emma has worked in various roles in consulting, research, and government (both state and Commonwealth). From 2007 to 2011, she worked in the Department of the Environment, Water, Heritage and the Arts, where she was responsible for scientific management issues to support the design and delivery of the Environmental Stewardship Program.

Anthea Coggan

Anthea Coggan is a research scientist/economist at CSIRO, specialising in transaction cost analysis and environmental offset policy design. Her work has identified the importance of identifying how the drivers of transaction cost interact with application context and stakeholder characteristics to determine the implications for policy.

Saul Cunningham

Saul Cunningham is a research scientist and team leader at CSIRO. His research projects have ranged from those focused on biodiversity conservation in fragmented systems, to those with a goal of lifting agricultural productivity, but he sees the biggest challenges at the intersection between conservation and production.

Graeme Doole

Graeme Doole is an Associate Professor in the Centre of Environmental Economics and Policy at the University of Western Australia, and Associate Professor at the University of Waikato in New Zealand. His key research interests involve the design of cost-effective programs to address the environmental impacts of dryland and temperate agricultural systems.

David Duncan

David Duncan works as a Research Fellow at the University of Melbourne, designing a strategy for quantitative evaluation of Australian Government environment programs. Previously, he worked as a senior scientist in government focusing on landscape and restoration ecology, native vegetation management, and monitoring and evaluation approaches.

Saan Ecker

Dr Saan Ecker is a consultant researcher drawing on the disciplines of human ecology, anthropology, ecology, and psychology. From 2008 to 2014, Saan was a senior scientist in the Australian Bureau of Agricultural and Resource Economics and Sciences social sciences research team, and led the team from 2010. Saan has 20 years' experience in natural resource management and sustainable agriculture, including management of multi-million dollar national resource management programs, development of regional plans and sustainable agriculture monitoring, reporting and change frameworks. His research topics include regenerative agriculture, environmental stewardship, farm diversification options, and a range of other topics relevant to regional Australia. His clients have included Australian and state governments and several national resource management groups.

Graham Fifield

Graham Fifield is a senior project manager at Greening Australia Capital Region, with seven years' experience in environmental rehabilitation. During this time, he has delivered a range of incentive funding projects on private and public land. Graham has worked extensively across the Southern Tablelands, South West Slopes and Central West regions of NSW. He estimates he has sown approximately 1,200 kms trees and shrubs using 400 kgs of native seed.

Robert Fraser

Rob Fraser is Professor of Agricultural Economics at the University of Kent, United Kingdom. He has an international research reputation as a policy economist, specialising in both agri-environmental and invasive species policy design and evaluation. He is a past President of the Agricultural Economics Society and is both a past President and a Distinguished Fellow of the Australian Agricultural and Resource Economics Society. He is also a member of the editorial board of the *Journal of Agricultural Economics*.

David Freudenberger

David Freudenberger is a lecturer and researcher in ecological restoration and management at The Australian National University. He has over 30 years of experience working in a diverse range of research areas, from grazing management to the impacts of landscape fragmentation. He was a senior scientist at CSIRO Wildlife and Ecology, and Chief Scientist at Greening Australia where he led many projects, including those on the effectiveness and cost of revegetation technologies and carbon sequestration measurement.

Philip Gibbons

Philip Gibbons is an Associate Professor at The Australian National University, with 25 years of experience in land management. He has worked as a park ranger, fire fighter, and forest ecologist, and currently plays a key role in forest management, native vegetation and biodiversity offset policies, and bush fire management.

Fiona Gibson

Fiona Gibson is Research Fellow at Centre for Environmental Economics and Policy at the University of Western Australia. She received her doctorate from the University of Western Australia in 2011. Fiona is currently working in the space of bushfire management, biodiversity, and water resources. Her research aim is to provide better advice to decision makers on effective policy design and the factors driving community adoption of such policies.

Romy Greiner

Romy Greiner is an environmental economist and Adjunct Professor at James Cook University. Her research has helped to uncover the underlying causes of and find workable solutions to a diverse range of sustainability challenges in regional and remote Australia, including biodiversity conservation, water management, salinity control, coastal and marine resource use. Romy is renowned for undertaking participatory action research with communities and nature-based industries — including traditional owners, graziers and pastoralists, farmers and cane growers, tourists and tourist operators, representative bodies, and national resource management groups. Romy's work has informed state and federal level policy development and her contributions support environmental management in northern Australia.

Md Sayed Iftekhar

Sayed is an environmental and resource economist with broad interests in the interactions between human and nature. He has received training on forestry (Khulna University) and biodiversity conservation (Oxford University), and worked on coastal zone management in Bangladesh for several years. He received his PhD from the University of Western Australia in 2012. He uses different economic tools, such as agent-based modelling, laboratory experiments, social survey, non-market valuation, and simulations, to study different environmental and natural resource management issues.

Geoffrey Kay

Geoffrey Kay is a research ecologist in the Fenner School of Environment and Society at The Australian National University. With over a decade of ecological field expertise, Geoffrey has a background founded in woodland ecology, biodiversity monitoring, and the conservation genetics of agricultural landscapes. His current research focuses on developing ways to advance the effectiveness of large-scale (trans-boundary) agri-environment conservation schemes.

David Lindenmayer

David Lindenmayer is Professor of Conservation Science in the Fenner School of Environment and Society, The Australian National University, and Science Director of the Australian Long Term Ecological Research Network. He has been working on long-term ecological research projects since 1983. He has published 38 books as well as over 950 scientific publications, which have addressed issues associated with ecological and biodiversity monitoring. Since 2009, David has managed the Environmental Stewardship Program's Box Gum Grassy Woodlands Monitoring Project. David Lindenmayer is a Fellow of the Australian Academy of Science and an ARC Laureate Fellow.

David Pannell

David Pannell is Professor and Head of School of Agricultural and Resource Economics at the University of Western Australia, Director of the Centre for Environmental Economics and Policy, Fellow of the Academy of Social Sciences in Australia, and was an ARC Federation Fellow from 2007 to 2012. David's research has won awards in the USA, Australia, Canada, and the UK.

Geoff Park

Geoff Park is a director of Natural Decisions Pty Ltd. From 1998 to 2013, he worked in a range of senior roles with the North Central Catchment Management Authority. From 2007, he worked as a knowledge broker, responsible for the development of collaborative partnerships between researchers, policymakers, extension staff, and landholders, which lead to improved knowledge exchange and on-ground biodiversity outcomes. Part of this involved working with a small research team exploring the development and application of the Investment Framework for Environmental Resources.

Maksym Polyakov

Maksym studied forestry and worked in forest management planning in Ukraine for number of years. He received a PhD in applied economics from Auburn University in 2004. He has held research positions at Auburn University, North Carolina State University, and the University of Western Australia, studying the economics of forestry harvesting behaviour, land use change, urban forestry, and ecological restoration. He is interested in the integration of ecology and economics to better understand the choices humans make concerning natural resources and the environmental consequences of these choices.

Paul Reich

Paul Reich is a research scientist at the Arthur Rylah Institute for Environmental Research. He works to understand and evaluate ecological responses to management activities, predominantly in aquatic systems, but more recently in native vegetation and threatened species management.

Anna Renwick

Anna Renwick is a postdoctoral fellow at the University of Queensland and a researcher in the ARC Centre of Excellence for Environmental Decisions. Her research focuses on looking at the trade-offs and synergies in ecosystem services, food security, and how to maximise biodiversity and social livelihoods. Her current projects include a national assessment of carbon, biodiversity, and co-benefits for indigenous people within the carbon farming initiative, investigating the effects of leakage on conservation planning, and determining the role of small-scale conservation efforts in agro-ecosystems.

David Salt

David Salt is the editor of *Decision Point*, the monthly research magazine of the ARC Centre of Excellence for Environmental Decisions. *Decision Point* presents news and views on environmental decision-making, biodiversity, and conservation planning and monitoring. Prior to working on *Decision Point*, David created and produced *The Helix* magazine for CSIRO Education, *Newton* magazine for Australian Geographic, *Materials Monthly* for ANU Centre for Science and Engineering of Materials, and *ScienceWise* for ANU College of Science.

Nancy Schellhorn

Nancy Schellhorn is a Principal Research Scientist with CSIRO. Her research focuses on landscape scale pest management. By combining large-scale experimentation with ecological modelling, she is seeking to inform landscape design and recommend management options for 'softening' the agricultural landscape matrix for the capture of ecosystem services of pest control.

Stuart Whitten

Stuart Whitten is a research scientist, economist, and group leader at CSIRO, specialising in the design and implementation of market-based instruments and other policy instruments to support environmental outcomes. He played a key role in the design of metrics to guide government investment in the Australian Government's Environmental Stewardship Program.

Charlie Zammit

Charlie Zammit holds a PhD in plant ecology and has spread his professional career between academia and government positions. From 2005 to 2012, he was Assistant Secretary of the Biodiversity Conservation Branch in the Department of the Environment, Water, Heritage and the Arts, where he was responsible for national biodiversity, vegetation, and forest policy issues, and for developing and implementing market-based approaches to biodiversity conservation on private land, including the Environmental Stewardship Program. He was part of the executive group for Caring for our Country. He retired in 2013 and is now an Adjunct Professor at the University of Queensland.

1

Introduction: Framing the agri-environment

Dean Ansell, Fiona Gibson, and David Salt

Conservation in an agricultural space

Do our agricultural landscapes hold the key to protecting our declining biodiversity? If they do, how would it be done? And who would pay? Would it be the landowner, or the general public (via the government)? These might sound like simple questions, but when you consider some of the environmental, social, and economic factors at play, it quickly becomes apparent that we are dealing with very complex issues.

To illustrate this, consider these two relatively simple situations, both examples of efforts to conserve biodiversity on farmland in Australia. The first involves a run-down paddock from which the landowner has removed his sheep and sown a mixture of native trees and shrubs in strips several metres apart. In exchange for a stewardship payment of $50 per hectare per year, the farmer agrees to keep his sheep out of the paddock for five years. He gets half the payment at the beginning and the rest at the end of the initial five-year period, at which time grazing stock are permitted back into the paddock under a regime where sheep are allowed into the site in short bursts (called 'pulse grazing') for the last five years of the agreement. By this time, the native vegetation should have developed enough to be able to cope with the reintroduction of grazing. Indeed, the presence of trees and shrubs will provide the grazing animals with valuable shelter.

Figure 1.1: Do our agricultural landscapes hold the key to protecting our declining biodiversity?
Source: Photo by Greening Australia.

The second situation involves a farmer agreeing to remove grazing sheep from a patch of box gum grassy woodland — an ecosystem now threatened in Australia. The farmer is allowed to let sheep into the woodland for pulse grazing, whereas previously the woodland experienced set stocking, meaning a certain number of animals were always there. The landowner also agreed not to use fertiliser in the woodland. For these actions, the government is prepared to pay the farmer over $200 per hectare per year, and the farmer has entered into a contract that will run for 15 years.

The first situation describes a process of restoration, with the aim of returning native vegetation to the landscape. It is about improving the natural value of degraded land, providing habitat for biodiversity and other environmental benefits. The second example is more about the preservation or conservation of an existing ecosystem. It is about sustaining the health and resilience of land with high natural values. Both schemes are undertaken in production landscapes, and the land under each scheme is expected to continue to provide agricultural outputs into the future.

Even with these simple descriptions, many questions immediately arise:

- Which approach is better for biodiversity, restoration, and/or conservation?
- Where do we get the best value for money? One farmer is paid four times the amount the other farmer receives; do we receive four times the return?
- Why should the government pay for a scheme which benefits the farmer (in the case of new trees providing shelter for stock)?
- Why does one scheme only run for 10 years when the other goes for 15?

Of course, there are many answers to each of these questions given by different groups. 'Which approach is better?', for example, would most likely be responded to differently by ecologists, economists, farmers, policymakers, and the public — and there would be considerable variation within each group. This variation simply underscores the complexity and uncertainty surrounding the operation of these schemes.

The two case studies described here are far from hypothetical exercises. They are based on real-life examples of publicly funded programs currently in operation on farmland in south eastern Australia. The first example (restoration) is called the Whole of Paddock Rehabilitation scheme (WOPR) being operated by Greening Australia (an environmental non-government organisation (eNGO)). The second case study (conservation) is part of an Australian Government program called the Environmental Stewardship Program. Both are described in more detail in this book (see Chapter 2 by Graham Field for background on WOPR, and Chapter 3 by Emma Burns and colleagues on the Environmental Stewardship Program).

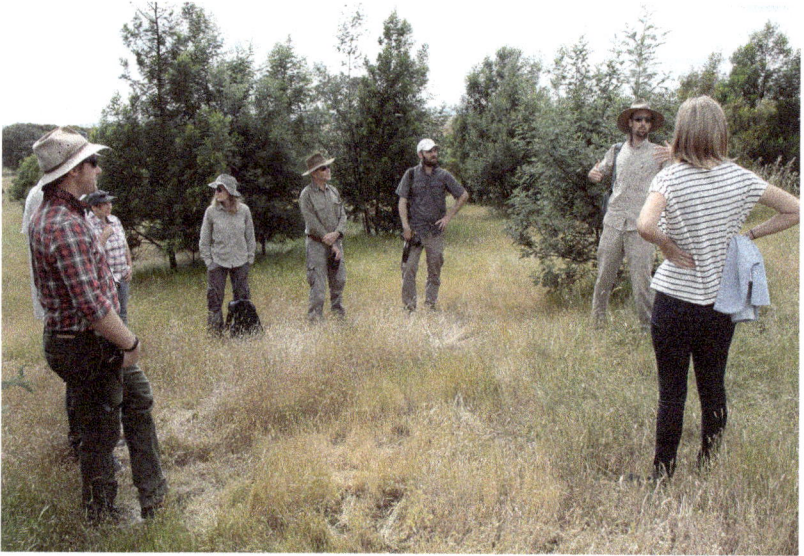

Figure 1.2: Agri-environment researchers and practitioners in a five-year-old WOPR site.
Source: Photo by David Salt.

In addition to having differing aims, payments, and duration, the schemes are also quite different in how they were developed and managed. WOPR came out of a grass-roots engagement between farmers and Greening Australia. The Environmental Stewardship Program was developed as a top-down government program to protect natural values that are considered to have national significance — in this case, the conservation of a threatened ecosystem. WOPR involved many 'back paddock' experiments, custom-made equipment, discussion, reflection, and trial and error (Streatfield et al. 2010). The Environmental Stewardship Program involved ecological, economic, and social science inputs, the development of legal contracts and the setting aside of funds beyond the traditional three- to four-year budget cycle.

WOPR and the Environmental Stewardship Program are but two examples of what are commonly referred to as 'agri-environment schemes'. There are many other variations of such schemes in Australia and around the world. Some, like WOPR, aim at restoring lost natural values. Others, like the Environmental Stewardship Program, aim to modify existing practice to conserve natural values.

We are not holding up these two schemes as examples of good or bad schemes. Rather, the differences between them offer a valuable reference point to discuss the strengths and weaknesses of society's effort to achieve environmental outcomes, generally regarded as public goods and services, from working agricultural landscapes, generally operated in the private realm. The particular environmental outcome this book focuses on is the conservation of biodiversity.

Before we begin to explore the many issues surrounding the design and implementation of effective agri-environment schemes, it is worth reflecting on the relationship between agriculture and biodiversity.

Why our farms are part of the solution

What is the connection between biodiversity conservation and our agricultural landscapes? Doesn't government look after biodiversity on behalf of the public through the creation and operation of national parks and nature reserves? Biodiversity conservation is an important goal of the management of most national parks, but the sad truth is that the world's system of nature reserves is not protecting biodiversity. A mere 15 per cent of threatened species on land are adequately covered by the existing network of reserves (Venter et al. 2014). In Australia, 80 per cent of threatened species are inadequately protected by the reserve system, with 12 per cent receiving no protection at all (Watson et al. 2010).

This is important because the world is witnessing a crisis of declining biodiversity. Species are being lost at 100–1,000 times what is believed to be the natural background rate of extinction, which scientists believe may have profound consequences for the future of human civilisation (Rockström et al. 2009). Governments around the world have signed up to the Convention on Biological Diversity, pledging that they will take actions that will slow and hopefully reverse these declines (Watson et al. 2014). To date, despite this commitment, little has been achieved. The fourth Global Biodiversity Outlook released by the United Nations in 2014 revealed that the rate of species loss is increasing and that the five principal drivers of extinction — habitat change, overexploitation, pollution, invasive species, and climate change — are getting worse (Secretariat of the Convention on Biological Diversity 2014).

Figure 1.3: An Environmental Stewardship Program site — a box gum grassy woodland in which grazing has been modified to protect the woodland's natural values.
Source: Photo by David Salt.

So what is the connection with farming? There are several broad areas to consider. The first relates to the point made above: our public reserve system is simply not providing adequate protection to our threatened biodiversity, as most threatened species and ecosystems lie outside of reserves, much of it on and around agricultural land. At least 40 per cent of global land surface is used for agriculture (Foley et al. 2005). In Australia, agriculture accounts for more than half of the land surface, with the majority of that land (86 per cent) used for grazing (Australian Bureau of Statistics 2014). If we want to conserve our biodiversity, we need to focus our efforts on agricultural land.

The second area relates to the impact of agriculture on biodiversity. About 70 per cent of the projected global loss of terrestrial biodiversity is attributed to agricultural drivers (Secretariat of the Convention on Biological Diversity 2014). The conversion of land to agriculture results in the loss and degradation of habitats. This directly impacts on plant and animal populations and communities, and alters ecological and hydrologic processes that underpin key ecosystem functions (Millennium Ecosystem Assessment 2005). Australia's settlement by Europeans over two centuries ago was followed by rapid and extensive landscape modification, as the settlers sought to tame the bush and establish grazing and cropping land. The initial focus was on clearing temperate grasslands and grassy woodlands (Kirkpatrick 1999). Records suggest that approximately half of the woody vegetation in Australia has been cleared since European settlement (Barson et al. 2000).

In addition to habitat loss, farming practices such as tillage, burning, livestock introduction, and nutrient and chemical usage have had significant negative impacts on biodiversity as well as soil, water, and air quality (Stoate et al. 2001). The early to mid-1900s saw a shift from smaller, low-input, mixed-enterprise farms to more intensive, specialised systems focusing on increased yields from fewer commodities, bringing with it increased fertiliser and pesticide use, and further loss of natural and man-made habitat (Bignal and McCracken 2000; Young et al. 2007). It is undeniable: agriculture has contributed greatly to the global decline in biodiversity.

The third area concerns the importance of biodiversity to the sustainability of our agricultural enterprises. Some elements of biodiversity underpin the quality and quantity of our agricultural output, through the provision of 'ecosystem services' — the wide range of benefits that we receive from ecosystems, ranging from food and water to recreation and cultural use. For example, bee pollination contributes more than €1 billion every year to Europe's strawberry producers (Klatt et al. 2014). Biodiversity can also provide benefits in the control of agricultural insect pests, improved soil fertility, and agricultural productivity (Altieri 1999). (The perceived importance of ecosystem goods and services to farmers is discussed by Saul Cunningham in Chapter 8.)

Lastly, as well as improving the financial outcome from farming, biodiversity benefits farmers by improving the amenity value of some properties and satisfying some farmers' goals of stewardship. (See Chapter 14 by Maksym Polyakov and David Pannell on the private benefits of biodiversity, and Chapter 12 by Saan Ecker and Chapter 13 by Romy Greiner on non-financial drivers of biodiversity conservation.) Biodiversity is also known to influence peoples' health and well-being (Keniger et al. 2013).

The bottom line is that agriculture requires the support of a raft of ecosystem services. The problem is that some of these services are valued more highly than others by agricultural producers, whose values may not align with those of the broader public. Juggling these contrasting values is one of the major challenges of farmland environmental management, and is a key theme throughout this book.

Solutions in the agri-environment

To anyone with an interest in conservation and agriculture, these ideas are hardly revolutionary. The question is, what can we do to conserve biodiversity in productive landscapes? There are many answers here, ranging from individual farmers volunteering their time and effort to re-establish native plants and animals on their farms, through to governments proclaiming laws regulating what farmers can and cannot do. As a spectrum of activity, these approaches might represent end points, going from volunteer effort through to regulation.

Most landowners have a limited capacity to sacrifice the productive capacity of their land (or their time) for non-income earning activities, and volunteer efforts have real limitations on what can be achieved (Curtis 2000). Indeed, the early investment in agri-environmental policy in the 1980s and 1990s focused on stimulating volunteer effort through programs such as Landcare. While popular, this effort failed to address the growing problems of land and water degradation and declining biodiversity (see Chapter 7 by David Salt).

Regulatory approaches, on the other hand, usually entail high transaction costs — especially, for example, in terms of compliance and enforcement — and are widely considered less efficient and cost-effective than alternative strategies (Hahn and Stavins 1992). They are also often unpopular in the agricultural sector. Indeed, the prevailing belief in most western democracies is that farmers have the implicit right to carry out the most profit-maximising activity on their land, irrespective of the external costs (and benefits) of doing so (Hanley et al. 1999). Regulation is usually only introduced where the activity is seen as being clearly unacceptable by the broader population, such as controlling the use of dangerous chemicals or the unacceptable treatment of livestock.

Between volunteering and regulation, however, there are many options employed and implemented by governments and conservation groups around the world, chief among which is the agri-environment scheme. Agri-environment schemes, though highly variable in their structure and application, can be broadly defined as programs involving payments to farmers in exchange for the provision of environmental goods and services (Burrell 2012; European Commission 2014). Most involve an acknowledgement that the farmer is sacrificing some aspect of their productive potential by providing environmental goods and services for the public good. The two case studies discussed at the beginning of this introductory chapter are examples of agri-environment schemes.

Over time, agri-environment schemes have attracted a growing share of government investment in agriculture across Organisation for Economic Co-operation and Development (OECD) countries, and now represent a significant component of biodiversity conservation in agricultural landscapes, with billions of dollars spent on such schemes around the world each year (Hajkowicz 2009).

Along with Europe and the United States, Australia has been working in the agri-environmental space for some 30 years. Australia's investment in this area has been tiny compared to Europe or the United States, partly reflecting our smaller population and economy, although the size of our agricultural landscape is comparable (Hajkowicz 2009). Given the enormous scale of the environmental challenges being faced in Australia, it is important that our investments in the agri-environment area are cost-effective.

Learning from agri-environment schemes in Australia: About this book

This book is targeted primarily at anyone working in agri-environmental policy or looking at establishing an agri-environment scheme in Australia, including policymakers, project officers, and non-government organisations. It has a secondary aim of producing a short and readable text for anyone interested in the topic of biodiversity conservation on agricultural land.

Chapters are short, engaging, and seek to educate rather than exhaustively prove finer points of analysis. Where possible, we have kept the use of jargon and acronyms to a minimum. Each chapter is a stand-alone story, and we have organised the book into the following three themed sections.

Part I — The agri-environment in the real world sets the scene by describing the challenges and tensions that go hand in hand with running agri-environment schemes. The chapters in this section present a variety of discussions of the complexities surrounding how agri-environment schemes function in real life, and discuss the case studies of the WOPR scheme and the Environmental Stewardship Program, which were presented at the beginning of this chapter. Part I also provides some contextual history of agri-environment schemes in Australia and Europe, and discusses dealing with different types of farmers and the importance of non-government organisations.

Part II — The birds and the beef explores the many natural, social, and economic values involved in agri-environment schemes, and the ways these are framed or marketed. In this section, we discuss the concept of ecosystem services, consider the debate over different conservation strategies, and are presented with an economics perspective on restoration. We also explore the issue of scale in designing agri-environment schemes, the importance of accounting for private benefits in project selection, and the social and psychological dimensions of agri-environment schemes.

Part III — Planning, doing and learning examines many of the issues surrounding the design, implementation, and evaluation of agri-environment schemes. It examines the many challenges of ranking different projects, given that most schemes are oversubscribed — there is never enough money to go around, so how do you get the best outcomes? We discuss approaches to measuring and maximising the conservation benefits, the importance of counterfactual thinking, and the choice of different policy tools. We conclude with the reflections of David Pannell, one of Australia's most experienced agricultural economists, on the performance of agri-environmental policies. He provides a checklist of factors that experience has shown are important to the success of any agri-environment scheme. For anyone with an interest or responsibility in agri-environment policy, this is one list you cannot afford to ignore.

So, what does it all this add up to? We attempt to make sense of the many perspectives in this book in the concluding chapter. We begin our conclusion with a simple hypothetical: if circumstances were to suddenly create a funding opportunity for a new agri-environment scheme, how should the nation respond? This is not idle speculation, because in many ways the Decade of Landcare was not an opportunity that was widely anticipated. It arose from a historic agreement between the National Farmers Federation and the Australian Conservation Foundation in 1989, coupled with a receptive prime minister. And while that scheme was enthusiastically embraced, it did not generate the enduring environmental outcomes that many hoped for.

A quarter of a century later, and the threat of environmental decline is as great, if not greater, with a rising expectation that our agricultural landscapes will dramatically increase their productive output in order to feed a growing population (see Box 1.1). Furthermore, biodiversity decline is just one of several issues facing society that must compete for limited funds.

Box 1.1: Farming, biodiversity and the future

The world's population is changing rapidly. In the next three decades there will be up to 10 billion people on the planet; Australia's population alone is expected to double by 2075 (Australian Bureau of Statistics 2013). Not only can we expect a lot more mouths to feed, but improvements to the socio-economic status of people across many regions, including Asia and Africa, will lead to changes in diet. This will result in a large increase in food demand, which will in turn require increased food production through the expansion and intensification of agriculture (Phalan, Green, and Balmford 2014). We will need to produce more with less.

The Australian Government's agricultural policy is heavily focused on capitalising on this growth by increasing productivity. The National Food Plan seeks to increase agricultural productivity by 30 per cent by 2025, aiming to increase the value of agricultural exports by 45 per cent (DAFF 2013). The Agricultural Competitiveness Green Paper sets out a plan to increase farm-gate profits by reducing costs and 'unnecessary barriers to productivity and profitability' (Commonwealth of Australia 2014). At the same time, this policy is aiming to 'streamline' the environmental approvals established through key legislation such as the *Environmental Protection and Biodiversity Conservation Act 1999*.

These major changes to agriculture present a significant threat to our biodiversity. Agricultural intensification carries greater biodiversity impacts than extensive farming practices (Reidsma et al. 2006). The amount of remnant vegetation expected to be cleared globally for agricultural use in the next 35 years is in the order of 0.2–1 billion hectares (Tilman et al. 2011). Facilitating the conservation of biodiversity in agricultural landscapes in the face of growing agricultural production represents a key conservation challenge at a global scale (Green et al. 2005). Policies that use incentives to balance conservation and agricultural production will play an increasingly vital role in safeguarding biodiversity in agricultural landscapes.

Should another major opportunity present itself — the announcement of a substantial government investment in agri-environment schemes, for example — will we be able to say we are ready? We should be, after 25 years of experience and research in these programs. Many of the perspectives in this book question our efforts in agri-environment investment and ask exactly what we have learnt. In many places it is suggested we can do a lot better than we currently do with the available resources, in areas including planning, prioritisation, monitoring, evaluation, and learning. In that light, it is our hope that this book will prove an invaluable resource and reference.

References

Altieri, M.A. (1999) 'The ecological role of biodiversity in agroecosystems', *Agriculture, Ecosystems and Environment* 74(1–3): 19–31. Available at: dx.doi.org/10.1016/S0167-8809(99)00028-6.

Australian Bureau of Statistics (2013) *Population projections: Australia*, Commonwealth of Australia, Canberra.

Australian Bureau of Statistics (2014) *Australian Environmental–Economic Accounts 2014*, Commonwealth of Australia, Canberra.

Barson, M., L. Randall, and V. Bordas (2000) *Land Cover Change in Australia: Results of the collaborative Bureau of Rural Sciences — State agencies' Project on Remote Sensing of Land Cover Change*, Bureau of Rural Sciences, Canberra.

Bignal, E.M. and D.I. McCracken (2000) 'The nature conservation value of European traditional farming systems', *Environmental Reviews* 8(3): 149–71. DOI:10.1139/er-8-3-149.

Burrell, A. (2012) 'Evaluating policies for delivering agri-environmental public goods', *Evaluation of agri-environmental policies: Selected methodological issues and case studies*, OECD Publishing, pp. 49–68.

Clayton, H., S. Dovers and P. Harris (2011) *HC Coombs Policy Forum NRM initiative*, The Australian National University, Canberra. Available at: crawford.anu.edu.au/public_policy_community/research/nrm/NRM_Ref_Group_Literature_review.pdf.

Curtis, A. (2000) 'Landcare: Approaching the limits of voluntary action', *Australian Journal of Environmental Management* 7: 19–27.

Commonwealth of Australia (2014) *Agricultural Competitiveness Green Paper*, Commonwealth of Australia, Canberra.

Department of Agriculture, Fisheries and Forestry (DAFF) (2013), *National Food Plan: Our food future*, Commonwealth of Australia, Canberra.

European Commission (2014) Agri-environment measures: Agriculture and rural development. Available at: ec.europa.eu/agriculture/envir/measures/index_en.htm.

Foley, J.A., R. Defries, G.P. Asner, et al. (2005) 'Global consequences of land use', *Science* 309: 570–4. DOI:10.1126/science.1111772.

Green, R.E., S.J. Cornell, J.P.W. Scharlemann and A. Balmford (2005) 'Farming and the fate of wild nature', *Science* 307(5709): 550–5. DOI:10.1126/science.1106049.

Hahn, R.W. and R.N. Stavins (1992) 'Economic incentives for environmental protection: Integrating theory and practice', *American Economic Review* 82(2): 464–8.

Hajkowicz, S. (2009) 'The evolution of Australia's natural resource management programs: Towards improved targeting and evaluation of investments', *Land Use Policy* 26: 471–8.

Hanley, N., M. Whitby and I. Simpson (1999) 'Assessing the success of agri-environmental policy in the UK', *Land Use Policy* 16: 67–80.

Keniger, L.E., K.J. Gaston, K.N. Irvine and R.A. Fuller (2013) 'What are the benefits of interacting with nature?', *International Journal of Environmental Research and Public Health* 10: 913–35.

Kirkpatrick, J. (1999) *A continent transformed: Human impact on the natural vegetation of Australia*, Oxford University Press, Melbourne.

Klatt, B.K., A. Holzschuh, C. Westphal, et al. (2014) 'Bee pollination improves crop quality, shelf life and commercial value', *Proceedings of the Royal Society B: Biological Sciences* 281: 20132440. DOI: 10.1098/rspb.2013.2440.

Millennium Ecosystem Assessment (2005) *Ecosystems and human well-being: Synthesis*, Island Press, Washington, DC. DOI:10.1088/1755-1307/6/3/432007.

Phalan, B., R. Green and A. Balmford (2014) 'Closing yield gaps: Perils and possibilities for biodiversity conservation', *Philosophical Transactions of the Royal Society B: Biological Sciences* 369(1639): 20120285. DOI:10.1098/rstb.2012.0285.

Reidsma, P., T. Tekelenburg, M. van den Berg and R. Alkemade (2006) 'Impacts of land-use change on biodiversity: An assessment of agricultural biodiversity in the European Union', *Agriculture, Ecosystems and Environment* 114(1): 86–102. DOI:10.1016/j.agee.2005.11.026.

Rockström, J., W. Steffen, K. Noone, et al. (2009) 'Planetary boundaries: Exploring the safe operating space for humanity', *Ecology and Society* 14(2): 32.

Secretariat of the Convention on Biological Diversity (2014) *Global Biodiversity Outlook 4*, Montréal. Available at: www.cbd.int/gbo/gbo4/publication/gbo4-en.pdf.

Stoate, C., N.D. Boatman, R.J. Borralho, C.R. Carvalho, G.R. de Snoo and P. Eden (2001) 'Ecological impacts of arable intensification in Europe', *Journal of Environmental Management* 63: 337–65. DOI:10.1006/jema.2001.0473.

Streatfield, S., G. Fifield and M. Pickup (2010) 'Whole of Paddock Rehabilitation (WOPR): A practical approach to restoring grassy box woodlands', *Temperate Woodland Conservation and Management* (eds D. Lindenmayer, A. Bennett and R. J. Hobbs), CSIRO Publishing, Melbourne, pp. 23–31.

Tilman, D., C. Balzer, J. Hill and B.L. Befort (2011) 'Global food demand and the sustainable intensification of agriculture', *Proceedings of the National Academy of Sciences* 108: 20260–4. DOI:10.1073/pnas.1116437108.

Venter, O., R.A. Fuller, D.B. Segan, et al. (2014) 'Targeting global protected area expansion for imperiled biodiversity', *PLoS Biology* 12(6): e1001891. DOI:10.1371/journal.pbio.1001891.

Watson, J.E.M., N. Dudley, D.B. Segan and M. Hockings (2014) 'The performance and potential of protected areas', *Nature* 515: 67–73.

Watson, J.E.M., M.C. Evans, J. Carwardine, et al. (2010) 'The capacity of Australia's protected-area system to represent threatened species', *Conservation Biology* 25(2): 324–32. DOI:10.1111/j.1523-1739.2010.01587.x.

Young, J., C. Richards, A. Fischer and L. Halada (2007) 'Conflicts between biodiversity conservation and human activities in the Central and Eastern European countries', *Ambio* 36(7): 545–50.

Part I.
The agri-environment in the real world

2

Working effectively with farmers on agri-environment investment

Graham Fifield

Key lessons

- Agricultural communities are now more diverse than ever, therefore incentive schemes must be flexible, and developed, at least in part, in consultation with the intended audience.

- Less than perfect ecological outcomes may be better than no outcomes at all.

- Voluntary schemes, whether subsidised or incentivised, deliver cost-effective outcomes, but must have ownership by landholders.

- Our largest scheme for private land revegetation was collaboratively developed with landholders and has uptake across the country.

- Environmental change takes time and requires an ongoing commitment to the site and the landholder to guarantee a return on the initial investment.

As a project manager with Greening Australia, I have been fortunate to work with a broad range of farmers and landowners on a variety of restoration and rehabilitation schemes in south east New South Wales and the Australian Capital Territory. This involves frequent collaboration with a range of scientists working in the agri-environment realm. (Indeed, I became involved in natural resource management (NRM) by undertaking an Honours Science degree in resources and environmental management at ANU). This experience has given me some overview of the ecological, economic, and social elements of the agri-environment.

Figure 2.1: Graham Fifield (left) listens to farmer Bob Spiller talking about his experience with Whole of Paddock Rehabilitation.

Source: Photo by David Salt.

It is my opinion that for any agri-environmental program to be effective it needs to acknowledge the diversity of the communities it engages with, and be implemented in such a way as to give ownership to farmers and encourage a voluntary ethic. Feedback from landowners and an understanding of the social barriers to adoption are critical when designing agri-environment schemes because without landholder support, schemes are unlikely to deliver the results desired by funding agencies — a view supported by Vanclay (2011). Furthermore, it is Greening Australia's experience that farmers often supply the innovation that will produce real and enduring results.

Agricultural communities are now more diverse than ever, therefore incentive schemes must be flexible and developed, at least in part, from the bottom up

Delivering any agri-environment scheme on private land requires an understanding of the agricultural enterprise, if one is being undertaken, and the social drivers for adoption. That is because these days the description of 'farmer' is an unhelpful stereotype that doesn't encompass the variety of people that are out on the land. The rural demographic is now as diverse as Australia's many unique and wonderful landscapes. Farms come in many forms, from the ever increasing lifestyle farms and peri-urban developments, to the dwindling numbers of traditional family farms as well as the increasing number of corporate agri-businesses. (See Barr 2009 for an overview of the changing agricultural landscape.)

Some of the reasons people get involved with revegetation or conservation programs are to provide shade and shelter for livestock, reduce the impact of soil erosion or salinity, increase birdlife on the farm, provide cleaner water for stock and fish, or to provide free fencing to assist management or as a buffer to adjacent land use. Other reasons have nothing to do with farming or the environment per se, such as for improved aesthetics, to increase property values or simply to block the view of neighbouring houses. Without delivering one or more products or services that the landholder values, we are unlikely to achieve repeat projects or widespread adoption. With this in mind, successful programs are typically those which are devised from the bottom up.

Are less than perfect ecological outcomes better than no outcomes at all? A case study in working with farmers

Herein lies a challenge: there is often a conflict between the best available science and the expectations of the project's participants. For example, revegetation guidelines for south east Australia (e.g. Munro and Lindenmayer 2011; Taws 2007) clearly state that to create habitat for declining or threatened woodland bird species, bigger is better, and revegetation in wider or square configurations is preferable to narrow linear strips. Many farmers, however, desire a long narrow configuration which acts as a windbreak over a large area of paddock, and the perceived loss of agricultural production caused by planting trees over more land is a serious concern. For smaller landholders, wider corridors can simply take up too much land proportional to their holding.

While a 15-metre wide native vegetation corridor with four rows of native trees and shrubs provides habitat for a range of birds, including many of conservation interest, a 25- or 30-metre wide corridor provides habitat for even more. But if we are uncompromising and insist on the best ecological outcome and the landholder isn't willing to forfeit the extra land, we risk achieving no outcomes at all. The question that arises then is: Are less than perfect ecological outcomes better than no outcomes at all?

If the demand for incentive schemes exceeds the capacity to supply, then, yes, we can prioritise towards the best environmental and most cost-effective outcomes. Over 30 years, however, Greening Australia has seen the benefit of working with willing and early adopters and getting early runs on the board. The value of a demonstration site that can be seen by others cannot be underestimated. To continue the wildlife corridor example, the benefits of having corridors on farms, such as improved stock shelter and increased birdlife, were demonstrated in an era where planting trees was considered radical. Incrementally, the next generation of corridors became progressively wider. The cost of fencing was the same regardless of width, so the desire to increase the shelter and wildlife benefits increased, and the loss of productive land became less of a concern. Locally, we have seen the transition from one- and two-row revegetation corridors in the early 1990s, to 12 metres wide, to 15 metres, and now it is generally accepted to create 18-, 20- or even 25-metre-wide corridors.

Figure 2.2: Over time, landholders have increased the size of their linear plantings.
Source: Photo by Dean Ansell.

Locally, the demand for these native vegetation corridors, even in the wider configurations, now exceeds our capacity to supply — it's a nice problem to have. I would suggest that if we had insisted on 20-metre wide corridors in the early 1990s, we would not have gained the necessary traction within the agricultural community to be in the position we are in now. There are still those landholders, however, often positioned in key locations for habitat connectivity, who insist on the current minimum width of 18 metres. With the trade-offs between the best available science and prioritising cost-effective actions described above, would you fund these?

In the case that supply exceeds demand for project funding, we typically seek to work with early adopters and innovators. These are often members of local Landcare groups. Further advertising may then be required to attract broader participation in a program. It is interesting to note that delivering projects in a region where Greening Australia has not had a strong or continuous presence often follows this path, highlighting the value in regional staff, regional offices, and a connection to regional areas.

Voluntary schemes, whether subsidised or incentivised, deliver cost-effective outcomes, but must have ownership by landholders

As a non-government organisation, Greening Australia is reliant on voluntary and incentivised schemes to achieve environmental outcomes on private land. Whilst Greening Australia does not have a legislative stick at its disposal, as agencies do, neither does it have to satisfy the same expectations regarding accountability. This has enabled Greening Australia to develop an incentive framework that appeals to landholders where ownership and personal investment (often sweat equity)[1] are required in the project, but the agreements are simple, concise, and not legally binding. The investment of time, money, and/or labour by the landholder is thus invested in their project and it is valued accordingly. It is our experience that, without this investment, environmental outcomes are easily compromised — stock enter exclusion areas, planted trees die, and electric fences stop working. In short, they have a stake in ensuring their project works.

Here are a few examples:

1. *The establishment of linear vegetation corridors.* Also known as windbreaks or wildlife corridors, the establishment of native vegetation in multiple rows along an existing fence line is now a feature of the agricultural landscape. Through corporate or government funds, Greening Australia disseminates a cash payment to the farmer to purchase the materials for the new fence to exclude stock. Importantly, he or she then builds the fence, or pays a contractor to complete the job. One of the best and longest running examples of this style of project is GreenGrid (a partnership between Transgrid and Greening Australia, see www.greeningaustralia.org.au/partner/transgrid). The new vegetation may be established by planting tubestock or by direct seeding. Tubestock are supplied free of charge from the Greening Australia nursery, but the landholder is responsible

---

1   An interest in a property earned by a tenant in return for labour towards upkeep or restoration.

for the site preparation and planting. Conversely, direct seeding is carried out by experienced Greening Australia staff while the landholder is responsible for site preparation. Their share of input to the project — whether it be labour, materials, time, or cash — is approximately 50 per cent.

2. *Protection of rivers, creeks or existing native vegetation.* Commonly known as fencing incentive programs, Greening Australia provides funds to the farmer to buy fencing materials. Again, the farmer is responsible for erecting the fence. In the case of rivers, payments towards alternate water sources, such as pipes and troughs, are available. The installation of this infrastructure is up to the farmer. Rivers in Australia are commonly invaded by woody weeds, and in the temperate zone these are typically willows (*Salix* sp.). Removal of mature willow trees is performed by contractors with heavy machinery and is funded through Greening Australia, with the smaller follow up infestation control performed by the landholder. Both parties have considerable input into the project. (See the Rivers of Carbon project at: riversofcarbon.org.au/.)

3. *Whole of Paddock Rehabilitation* (WOPR, pronounced 'whopper' — see Fifield et al. 2014). In 2008, Greening Australia launched a new style of incentive scheme that combines traditional revegetation incentives, such as fencing and revegetation, with fixed-term stewardship payments to offset the loss of agricultural production during a short period in which native vegetation is established and matures. The payment has been deliberately set at a rate that is less than the full productive potential of agriculture (typically 30–60 per cent, depending on soil type) thus requiring an additional investment by the farmer into the project. Once trees and shrubs are sufficiently established, typically after five years, agricultural production resumes and the stewardship payments are stopped. Direct seeding and fencing (where occasionally required) follow the models described above. Approximately 3,000 hectares of productive agricultural land within the threatened Grassy Box Woodland communities of temperate Australia have now been revegetated and productively enhanced using this approach. WOPR is also a great example of innovation. Passive tree regeneration schemes with similar payments to landowners exist in Victoria (the Bush Returns Project described by Miles 2008).

Our largest scheme for private land revegetation was devised by landholders and now has uptake across the country

Farmers are fantastic innovators; working with them and respecting their knowledge and skills can result in excellent outcomes. A good example of such collaboration can be seen in WOPR.

WOPR grew out of an experiment between a couple of local farmers and Greening Australia in south east NSW. In 1989, following bushfires and an alarming rise in the water table, a paddock on one of these farmer's properties became waterlogged, salty, scalded, and provided little agricultural productivity. The landholder contacted Greening Australia and the phenomenon was identified as dryland salinity.

In 1994, the farmer decided to sow native trees and shrubs across this paddock to combat the high water table. The design was to sow alleys of trees, comprising four rows of trees and shrubs separated by 30–50 metres of pasture for grazing between each alley. The wisdom of this became clear after four or five years when sheep were reintroduced to graze amongst the trees. The widely spaced tree alleys didn't suppress grass growth over large areas of the paddock, while providing shade, shelter, additional fodder, and lowering the water table. Anecdotally, the rest from grazing benefited the native perennial grasses, which were able to grow unhindered and set seed. Areas that were bare and scalded stabilised. The diversity of trees and shrubs provided valuable habitat, and birds that hadn't been seen on the property for many years returned to the area.

WOPR is essentially a form of alley farming, as described by Lefroy and Stirzaker (1997). Greening Australia has refined the scheme over several years to include stewardship payments, and by adjusting the design and seeding rates to strive for the best social, agricultural, and ecological outcomes. Several iterations of the program have been undertaken with money from a variety of sources. In every case, the scheme was readily accepted by farmers because it was easy for them to incorporate into their farm budget. Paddocks were a unit of production, Greening Australia was a trusted partner, and the scheme did not involve legal contracts or involve heavy transaction costs. WOPR agreements simply give the farmers half of the stewardship payment up front and half at five years, if the farmer has met their side of the bargain — not putting sheep in. Since the project's inception in 2008, approximately 85 per cent of landholders have qualified for the second stewardship payment. The remaining 15 per cent of projects have typically been cancelled due to poor plant establishment rather than deliberate stock grazing. It is worth noting that when properties have changed owners, the second stewardship payment, which is available to the new owners, has provided the incentive to continue with the stock-free period and thus contributed to the success of the project.

Figure 2.3: A before and after photo of a WOPR project near Bookham, NSW.
Source: Photo by Graham Fifield.

WOPR is now being implemented at a larger scale, with Greening Australia receiving significant grants from the federal government's Caring for Our Country (CfoC) program in 2011 and the Biodiversity Fund in 2013 to run region-wide programs across several catchments in south east NSW. A variation of the program is now being delivered in the Avon valley of Western Australia and is being considered across several states and territories of Australia.

Environmental change takes time and requires an ongoing commitment to the site and the landholder to guarantee a return on the initial investment

Greening Australia has planted many trees in its time and delivered many great environmental outcomes. (For examples, see Briggs et al. 2008; Spooner et al. 2002 — remnant fencing; Gould 2013; Higgisson 2014 — riparian restoration; Taws 2007; Lindenmayer et al. 2012 — birds in revegetation; Gibson-Roy et al. 2010 — grassland restoration). But possibly a greater impact has been achieved in the relationships and knowledge it has established in the farming community. Restoration takes time, trees grow slowly, so it's important to nurture long-lasting relationships with the landowners who are participating in the various agri-environmental schemes being run around the country. Unfortunately, it is not enough to simply throw a few handfuls of seed over the land and walk away. There will be many challenges facing the landowner with the project over the years, so follow up and being available to provide advice is essential.

A major issue with some government agri-environment programs is that once the program has concluded — and they rarely run longer than a few years — landowners don't have anyone to provide them with advice or feedback on what they should do when problems arise. This may be exacerbated by the loss of NRM staff to state agencies (Curtis et al. 2014). This is where non-government organisations such as Greening Australia and various catchment management groups play an essential role (see Chapter 5 by David Freudenberger). Follow up visits are a challenge, as they are never directly funded, but rather have to be conducted while delivering other projects or from organisational surpluses.

Issues that may arise with project sites are, of course, varied and may include such issues as 'my direct seeding hasn't worked', 'the erosion is still occurring', or 'what should I do about this troublesome animal/ plant?'. Without subsequent advice, it is possible that livestock may be allowed to enter the site and graze on the tiny seedlings, that the erosion continues to occur, or that a noxious plant or animal compromises the success of the project. In each case, the investment of public and private time, money, and resources is jeopardised.

Of course, this ongoing relationship is essential for Greening Australia too, and is a necessary investment. One of the challenges in environmental restoration is predicting the trajectory of a site in five, 10 or 100 years into the future. It is only by carefully recording what actions are undertaken today and checking what the site looks like tomorrow that we can begin to learn and adapt our methodology for the best social, ecological, and agricultural outcomes.

Acknowledgements

I need to acknowledge my colleagues at Greening Australia, past and present, for developing and nurturing the programs and culture that persists to this day. These include, but are not limited to, Bindi Vanzella, Angela Calliess, Lori Gould, Sue Streatfield, Nicki Taws, and Brian Cumberland.

We are indebted to many local landholders, in particular Leon Garry and John Weatherstone, for their innovative paddock-scale solutions to paddock-scale problems.

Thanks to the detailed comments provided by reviewers David Freudenberger and David Duncan.

References

Barr, N. (2009) *The house of the hill: The transformation of Australia's farming communities*, Land and Water Australia, Canberra.

Briggs, S.V., N.M. Taws, J.A. Seddon and V. Vanzella (2008) 'Condition of fenced and unfenced remnant vegetation in inland catchments in south-eastern Australia', *Australian Journal of Botany* 56(7): 590–9.

Curtis, A., H. Ross, G.R. Marshall, et al. (2014) 'The great experiment with devolved NRM governance: Lessons from community engagement in Australia and New Zealand since the 1980s', *Australasian Journal of Environmental Management* 21(2): 175–99.

Fifield, G., S. Streatfield and C. Ross (2014) *Introducing Whole of Paddock Rehabilitation: A new approach to regreening the farm*, Greening Australia, Canberra. Available at: www.greeningaustralia.org.au/ uploads/knowledge-portal/ACT_WOPR_brochure_2014.pdf.

Gibson-Roy, P., G. Moore, J. Delpratt and J. Gardner (2010) 'Expanding horizons for herbaceous ecosystem restoration: The Grassy Groundcover Restoration Project', *Ecological Management and Restoration* 11(3): 176–86.

Gould, L. (2013) *Analytical case study of a large scale riparian rehabilitation project from an NRM perspective: Boorowa River Recovery*, Masters final report. Available at: www.riverfoundation. org.au/admin/multipart_forms/mpf__resource_310_8___ Boorowa%20RR%20Evaluation%20Gould%202013.pdf.

Higgisson, W. (2014) *An evaluation of riparian restoration: A case study from the Upper Murrumbidgee Catchment, NSW, Australia*, Honours thesis, Institute for Applied Ecology, University of Canberra.

Lefroy, E.C. and R. Stirzaker (1997) *Alley farming in Australia: Current research and future directions*, CSIRO Centre for Environmental Mechanics, Canberra.

Lindenmayer, D.B., A.R. Northrop-Mackie, R. Montague-Drake, et al. (2012) 'Not all kinds of revegetation are created equal: Revegetation type influences bird assemblages in threatened Australian woodland ecosystems', *PLOS ONE* 7(4): e34527. DOI:10.1371/ journal.pone.0034527.

Lovett, S. and G. Lori (2015) *Rivers of Carbon*. Available at: riversofcarbon.org.au/.

Miles, C. (2008) 'Testing market based instruments for conservation in northern Victoria', *Integrating Conservation and Production: Case Studies from Australian Farms, Forests and Fisheries* (eds T. Lefroy, T. Bailey, G. Unwin and T. Norton), CSIRO Publishing, Canberra, pp. 133–46.

Munro, N. and D. Lindenmayer D. (2011) *Planting for Wildlife: A practical guide to restoring native woodlands*, CSIRO Publishing, Canberra.

Spooner, P., I. Lunt and W. Robinson (2002) 'Is fencing enough?: The short-term effects of stock exclusion in remnant grassy woodlands in southern NSW'. *Ecological Management and Restoration* 3(2): 117–26.

Taws, N. (2007) *Bringing back birds: A glovebox guide*, Greening Australia, Canberra.

Vanclay, F. (2011) 'Social Principles for agricultural extension in facilitating the adoption of new practices', *Changing land management: Adoption of new practices by rural landholders* (eds D. Pannell and F. Vanclay), CSIRO Publishing, Canberra.

3

The Environmental Stewardship Program: Lessons on creating long-term agri-environment schemes

Emma Burns, Charlie Zammit,
Simon Attwood and David Lindenmayer

Key lessons

The conservation of biodiversity on private land is both a high priority and a considerable challenge. An effective response to this challenge requires a combination of legislative and incentive mechanisms, coupled with preparedness by government to review and revise administrative arrangements. Preliminary results from the Environmental Stewardship Program, established by the Australian Government, highlight that there is a role for market-based approaches. However, implementation of this program through a Commonwealth bureaucracy was not without its challenges. Here we provide an overview of the program's implementation from 2007 to 2012, followed by discussion of some key lessons learned.

We summarise these lessons as:

- Designing for the long-term presents many challenges.
- Land managers liked the program but there were a few surprises.

- Monitoring is important.
- Start simple and engage early and often.
- Governance and administrative reforms are needed.

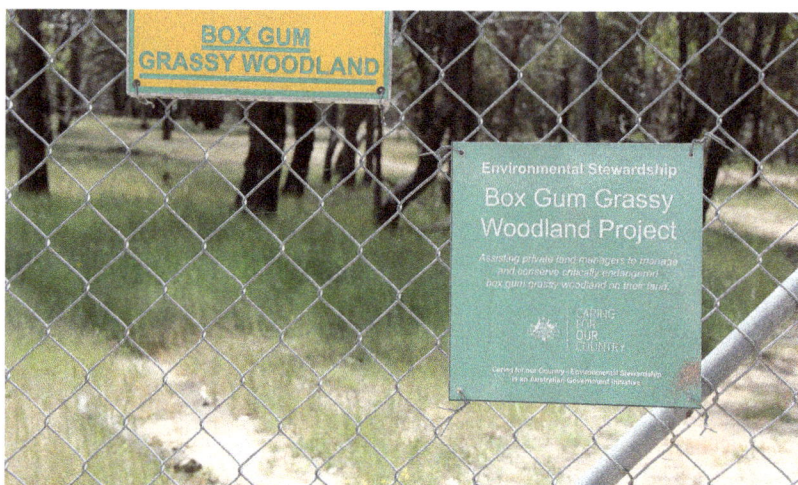

Figure 3.1: A sign on the gate of a property involved in the Box Gum Grassy Woodland Project, part of the Environmental Stewardship Program.
Source: Photo by David Salt.

Introducing the Environmental Stewardship Program

In mid-2007, the Commonwealth established the Environmental Stewardship Program, a ground-breaking scheme that used competitive tenders to engage private land managers in long-term contracts (up to 15 years) to manage environmental assets of high public value. The program resulted in a series of tenders being implemented by the Commonwealth across New South Wales, Queensland, and South Australia. We were involved in the design and implementation of the program and, more than most, we are aware of the challenging ecological, social, and economic dimensions of designing and implementing such a process. Here we reflect on the experience and offer several lessons that may help with the design of similar schemes in the future.

The Environmental Stewardship Program initiated a new way for the Commonwealth to support the conservation of biodiversity on private land, through a process where land managers were empowered and funded to be recognised as environmental stewards.

Being a Commonwealth initiative meant funding was targeted at matters of National Environmental Significance as listed under the *Environment Protection and Biodiversity Conservation Act 1999* (EPBC).[1] Therefore, the program was only permitted to target nationally threatened species and ecological communities, migratory species and wetlands of international importance, and natural values associated with world and national heritage places.

Aligning a market-based incentive scheme to clear Commonwealth legal responsibilities for biodiversity conservation was critical to gaining initial approval for the program. Depending on the assets targeted, the program sought to achieve a range of outcomes including:

- Improved habitat quality across the landscape.
- Increased viability, integrity, and buffers to high quality remnants for species, ecological communities, Ramsar wetlands, and World Heritage areas.
- Improved long-term protection of nationally threatened species and ecological communities.
- Improved condition and function of ecological communities.
- Enduring changes in land manager attitudes and behaviours towards environmental protection and sustainable land management practices.

The initial funding for the program was $42.5 million from 2007/08 to 2010/11, with a contingency reserve to allow annual payments until 2024/25 (a contingency reserve for a program represents funds committed for the program beyond the standard three-year forward estimates period). In the 2011 budget, the Commonwealth announced additional funding of $84.2 million from 2011/12 for a further four years. However, no further funding rounds were offered after 2012.

---

1 Note, not all departmental programs need to target matters of National Environmental Significance. They can have their constitutional basis through external affairs powers — helping the Australian Government meet their international obligations under the Convention on Biological Diversity, for example.

In the implementation phase of the program (2007–2012), managers designed and delivered the competitive tenders through two projects in collaboration with on-ground delivery agents and external scientific experts: the Box Gum Grassy Woodland Project and the Multiple Ecological Communities Project. Both projects, which comprised the entire program, employed a reverse auction tender process (see Zammit et al. 2010), which resulted in a total government investment of approximately $152 million in approved grants with individual land managers up to 2025/26. Landowner contributions remain uncosted, but are likely to be significant.

From 2007–2009, the program targeted the critically endangered box gum grassy woodland ecological community in south east Australia through the Box Gum Grassy Woodland Project. This project targeted the remaining patches of woodland on private land, without specific reference to the adjoining matrix of agricultural lands or other non-target native habitats. In total, five tender rounds across New South Wales and Queensland were conducted under the Box Gum Grassy Woodland Project, resulting in 26,470 ha being managed by 210 land managers for an approximate cost of $71 million over 15 years.

Program managers recognised an opportunity to increase the program's efficiency through experience gained from implementing the Box Gum Grassy Woodland Project; desktop research; staff expertise (see, for example, Attwood et al. 2009); and formal review and structured feedback mechanisms with delivery agents and land managers.

Consequently, they sought to improve program design by broadening the program's reach through targeting multiple EPBC-listed ecological communities in a region, and incorporating options for conservation management of the surrounding matrix through buffers and connectivity. These program refinements were subsequently found to have improved the efficiency and effectiveness of the implementation model (see, for example, Marsden Jacobs Associates 2010).

In 2010–2011, the program implemented the Multiple Ecological Communities Project in New South Wales and South Australia, across six different Natural Resource Management Regions. Five nationally threatened ecological communities were targeted: in New South Wales, basalt and alluvial grassland, weeping Myall woodland, and box gum grassy woodland; in South Australia, iron-grass grassland, and peppermint box woodland. In 2011–2012, a second round of the Multiple Ecological Communities Project was implemented in South Australia. In total, after these two tender rounds, 87 land managers were contracted to manage 26,988 ha of threatened ecological communities, which included over 7,000 ha of adjoining land for an approximate cost of $81.3 million over 15 years.

Relative to the Box Gum Grassy Woodland Project, the development of the Multiple Ecological Communities Project recognised the need for a more integrated and landscape-scale approach to conserving threatened ecological communities. As such, more technically nuanced protocols and tools were required (Whitten et al. 2011). These, in turn, required more sophisticated management planning with land managers. Building on the successful uptake of the Box Gum Grassy Woodland Project and Multiple Ecological Communities Project, the program managers commenced designing a more generic reverse auction framework that targeted native vegetation (habitat), which supported nationally threatened species and ecological communities. Under this approach, the program could then be rolled out without specifying a precise target, but rather allowing land managers with different assemblages of EPBC-listed species and communities on their properties to participate in a tender round. This approach was never implemented, given the Commonwealth's decision not to undertake further funding rounds of the program.

In summary, after five years of implementation, 297 land managers across New South Wales, Queensland, and South Australia were approved by the Commonwealth to implement (up to) 15-year conservation management plans over 56,527 ha of private land. The last of these contracts will end in 2026/27. The realisation of the potential conservation benefits from a public investment of approximately $152.3 million will depend on how these contracts are managed, and how land managers are supported.

Figure 3.2: A reptile monitoring station within a Box Gum Grassy Woodland Project site.
Source: Photo by David Salt.

The program can justifiably be called groundbreaking because it required a longer-term perspective to the management of grants than has customarily been taken by Commonwealth governments. Consequently, the program contained many innovations, which were developed and implemented in a relatively short period of time. As we look back on what was achieved, we believe there are several important lessons for policymakers seeking to set up similar schemes in the future.

Designing for the long-term presents many challenges

The original design of the program that supported long term payments for conservation management on private land was a significant achievement. The original budget — for the full funding term to 2026 — contained the necessary allocations for outsourcing the management of tenders to third-party providers, environmental and social monitoring, compliance, and extension support.

Retaining these features adequately through the implementation and maintenance phases of the program proved to be difficult due to a combination of factors, including budget pressures, changes in department staff, and changes in priorities and attitudes within government and the department towards how the program should best be managed. For example, the Box Gum Grassy Woodland Project featured ecological and social monitoring, and externally contracted delivery agents to manage site assessments and provided ongoing extension support to land managers. However, the Multiple Ecological Communities Project had few of these features. The challenge was presented by the fact that governments always retain the prerogative to reallocate limited funds, and other resources — and to shift priorities as circumstances change.

The translation of policy decisions into programs with long-term budgets can be difficult to maintain successfully over time. Original planning cannot deliver intended results without the institutional commitment and enduring support to implement a program as intended. This is a reflection of the vulnerability of agreed government policies and investment programs to shifting political ideologies and their preferences. Such shifts can limit long-term policy coherence in favour of short-term flexibility. This can, in turn, limit opportunities for securing enduring long-term environmental improvements. The major challenge for any future agri-environmental program will therefore be securing enduring bi-partisan political support, combined with institutional governance arrangements that make it more robust to withstand short-term pressures and shifting attitudes.

Land managers liked the program but there were a few surprises

Land manager feedback

The program was popular with land managers (Coggan et al. 2013; Ecker and Thompson 2010; Marsden Jacobs Associates 2010). The features of the program prompting the most positive feedback from land managers were site assessments and ongoing monitoring, information packs and evenings, and the use of state-and-transition models, which were used to explain the desired conservation outcomes. Zammit et al. (2010) and Attwood and Burns (2012) provide further information on the use of state-and-transition models by the program. Zammit (2013a) outlines the social benefits to farmers from participating in incentive programs for conservation. Here we focus on land manager feedback, with selected quotes from land managers from semi-structured interviews.

To some, the offer of 15-year contracts was appealing; to others it was daunting. Some found the Commonwealth's interest (through weighting in the metric) in conservation covenants — deeds to land titles that define the limitations, conditions, or restrictions on the use of that land in perpetuity (see www.environment.gov. au/topics/biodiversity/biodiversity-conservation/conservation-covenants) — was a barrier to participation, even though it was not a requirement to participate.

Many land managers were proud of their involvement and thought the department should go further to develop a brand for the program (something akin to current organic farming branding). The department did provide them with signs for display on properties (e.g. gates) that recognised their participation in the program, but did not develop a brand. As yet, there is no evidence of a clear market advantage to properties that have participated in such schemes, but as such schemes mature, the competitive advantages of products that arise from participating properties might be more evident:

> I would like to see stewardship branded as a premium product. We have happy sheep and look after the environment — wouldn't you want to buy that wool?

Some saw the reverse auction process as confusing and undesirable:

> I had a hard time coming up with a bid price because I didn't know what I was doing. Why don't you just tell us a flat price then we can decide if it is worth it?

At the outset, managers did not know a reasonable price to make direct offers, but after running multiple rounds within a region, there was sufficient price information for direct offers. This approach had already been successfully used in the Commonwealth's Tasmanian Forest Conservation Fund (Binney and Zammit 2010).

Some felt the management plans developed were too prescriptive and should be outcome-based rather than input-focused:

> Rather than tell us what to do, you should have a hands-on person come around and pay us a bonus if we are getting the outcome you want.

A few surprises

Implementation of the program produced some surprises, including a large number of requests for site assessments. For example, the Box Gum Grassy Woodland Project had initially budgeted for 150 site assessments, but received over 350 requests in round one. Project managers were further surprised by some applications worth millions of dollars over the contract period. To address this, managers introduced a capped total bid amount (e.g. $3.5 million maximum over 15 years for the Multiple Ecological Communities Project).

From initial rounds, project managers discovered that land managers were generally costing their bids linearly. As efficiencies were expected, this made large holdings more expensive than anticipated. As the program was rolled out, concern grew within the department regarding what was an appropriate $/ha/year figure for management. The concern was that, in many cases, it could be cheaper to purchase the property, as was the approach for funding the purchase and covenanting of private land through a state government or private entity under the National Reserve System. Under that approach, the new owner carries the ongoing costs to implement the conservation management plan.

In response to this concern, the Australian Government's Evaluation Panel (which was responsible for overseeing the process and recommending successful tenderers to ministers) developed a $/ha/year cap as a red flag when evaluating Multiple Ecological Communities Project bids. This figure was informed by cost data from existing Box Gum Grassy Woodland Project contracts, such as the average annual cost and the variation from the mean. This information is commercial-in-confidence. However, in a commissioned review, Marsden Jacobs Associates (2010) reported that across the Box Gum Grassy Woodland Project, the average annual cost was $202 per hectare per year, with significant variation around the average both within and between regions. The actual figure used was not communicated to land managers, but the Evaluation Panel's discretionary powers to support their responsibility to make the best value for money judgements was communicated in the program guidelines (see, for example, nrmonline.nrm.gov.au/catalog/mql:2408).

Monitoring is important

In recognising that several kinds of monitoring and engagement activities are needed, a set of monitoring tools and approaches were developed for the program. These aimed to:

- Provide feedback to land managers to engage them to increase their understanding of the program and its aims, and to engender positive attitudes towards the environment.
- Provide information for compliance checking, risk management, grant acquittal requirements, and departmental reporting for the program.
- Provide rigorous evidence for the performance of the program in achieving its conservation and attitudinal change outcomes.

A structured approach should highlight the benefits from the investment, and the positive behaviours and attitudes of those participating. Critical components include annual (short term) compliance reporting by land managers against contracted obligations; longer term ecological monitoring to reveal ecological improvements; and longer term social monitoring to track changes in attitudes and priorities to biodiversity conservation among farmers.

These monitoring systems are coupled with extension support, which gives land managers somewhere to go for advice, and provides them with opportunities to build capacity and share their learnings with other land managers, researchers and government. Finally, monitoring systems require regular independent auditing of a proportion of contracts to help detect and deal with fraudulent activity at an early stage.

The lack of fit-for-purpose long-term biodiversity monitoring has been a source of considerable criticism, from both scientific and policy perspectives, of agri-environment schemes in Europe, as well as earlier conservation initiatives in Australia (Kleijn and Sutherland 2003; Morrison et al. 2010; Lindenmayer et al. 2012). In recognising these shortcomings, the program contracted The Australian National University to undertake scientific monitoring of woodland sites on 153 farms in New South Wales and southern Queensland (Lindenmayer et al. 2012). Securing this monitoring project was a significant policy achievement, and was vital for determining progress towards achieving the program objective and desired outcomes.

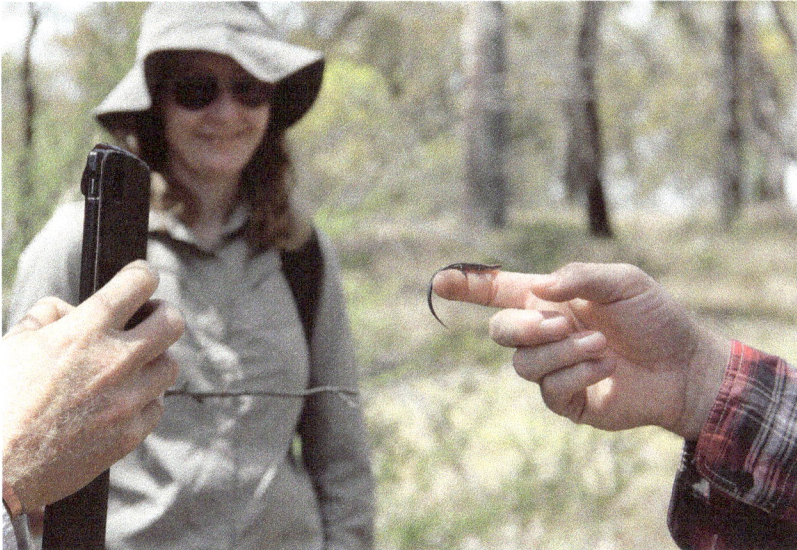

Figure 3.3: A native skink found in a monitoring station in a Box Gum Grassy Woodland Project site.
Source: Photo by David Salt.

The monitoring project had an initial budget of $2 million over four years. In 2013, the project was re-funded, but at half its initial budget. Consequently, the size of the monitoring has been adjusted by introducing rotational sampling, as described in Lindenmayer et al. (2012). In addition, some sites are no longer visited, such as those for which property ownership has changed and stewardship payments are no longer being made. Despite these changes, the monitoring project is still considered appropriate to assess many aspects of change in certain groups of biota, and woodland condition.

The results of the monitoring project are provided to the department in annual reports. These reports include evidence-based recommendations for alterations to the prescribed Box Gum Grassy Woodland Project grazing regime. To date, results from the monitoring indicate that the areas of vegetation being managed are on a different trajectory to the controls (currently being prepared for publication). More time is needed to understand these trajectories, but early indications are that the management plans are effective in achieving conservation outcomes. (See Kay et al. 2013 for an insight into reptile and amphibian assemblages at these sites.)

Start simple and engage early and often

When seeking to establish a new biodiversity market, it makes sense to begin with a simple investment target, but one that is sufficiently recognisable and widespread to ensure a reasonable number of eligible participants. In the case of the Environmental Stewardship Program, the targeted asset was box gum grassy woodlands. This critically endangered ecosystem is distinctive and widely spread from north eastern Victoria across western NSW and into southern Queensland.

Building on the targeting of a single asset (i.e. box gum grassy woodlands), one can then use early learnings to improve effectiveness and efficiency through *increasing* the amount of land secured and *decreasing* the administrative costs associated with program management. In the case of the Environmental Stewardship Program, this led to introducing multiple ecological communities into the reverse-auction process, and thereafter to the design of a more generic, habitat-based approach to tender design. The habitat-based approach leads logically into more explicit consideration of how landscape-scale outcomes can be secured through formal inclusion of opportunities for creating corridors and other kinds of functional connectivity (see Chapter 11 for Geoff Kay's discussion on how the landscape-scale of an agri-environment program can affect outcomes).

Agri-environment schemes and markets for biodiversity are a novel idea to many land managers, so misinterpretation of the process and resulting contracts is a risk. It is important to investigate concerns early and to regularly engage with land managers. It is also important to use simple tools to communicate program intentions, operational guidelines, and contractual conditions. As understanding grows and early successes become evident within rural communities, there is significant opportunity to build land manager capabilities for biodiversity conservation on their properties and to cultivate new social networks around these environmental assets. This is because some of the biophysical and social benefits from stewardship projects could be privately captured (i.e. be of personal rather than public benefit). This should help facilitate the acceptance of such programs within land manager communities (see Chapter 10 on restoring ecosystem services on private farmland).

Governance and administrative reforms are needed

Several independent reviews provided valuable insights and generally concluded that the program was well designed (Ecker and Thompson 2010; Marsden Jacob Associates 2010; Whitten et al. 2011). During this time, it was also well regarded by the ministers for the environment, the scientific community, many farmers, and the National Farmers Federation. The program proved successful in targeting threatened ecological communities, and in highlighting the role played by other native habitats and the surrounding agricultural matrix in market-based conservation management on private land.

There were other benefits for the department in meeting its legislative requirements under EPBC, including improved knowledge of the location, condition, and extent of certain threatened ecological communities, with flow on benefits to recovery planning processes. Most recently, a senate committee inquiry into threatened species protection endorsed the Environmental Stewardship Program, and recommended longer time frames for funding (see recommendations 25 and 32 of the senate committee report at: www.aph.gov.au/ Parliamentary_Business/Committees/Senate/Environment_and_ Communications/Completed_inquiries/2010-13/threatenedspecies/ index).

For all these benefits, however, there were also considerable challenges, and the Commonwealth closed the program to further investment rounds in 2012. While the authors do not know the reasons why the program ceased further investment after two successful projects, we speculate several factors had some influence on that decision. As a relatively small and discrete investment, the program was vulnerable to cost-savings efforts during the more restrictive fiscal budgets from 2012. Second, the program never secured the political support of the mainstream environmental non-government organisations during its life, so no pressure was placed on government by them when funding was under threat.

The program, as originally approved, was designed to be innovative in addressing the long-term management of specific matters of national environmental significance on private land. But the realities of implementing an innovative conservation program through a Commonwealth bureaucracy, with rigid business processes, were broadly underestimated. Essentially, implementing reverse auctions was demanding because running a cycle of market-tenders inside a broader culture based around funding open-call public grants caused a range of procedural and time-critical constraints.

There was also the issue of dealing with scientific knowledge and its application, something essential in designing a program aimed at sustaining complex threatened ecosystems. The Commonwealth successfully administers a number of technically and scientifically complex policy areas (e.g. the Bureau of Meteorology, Great Barrier Reef Marine Park Authority, and the Antarctic Division within the Department of the Environment) which are resourced and willing to support scientific research and analysis in policy design and program management. However, this kind of approach to scientific knowledge and its application was not fully adopted for the Environmental Stewardship Program. Rather, it was delivered through the Caring for our Country (CfoC) initiative, and administered by a bureaucracy with a primary focus on managing a large and complex national grants program. (CfoC involved spending $2 billion over four years.) However, an administrative culture more akin to those in the above Commonwealth areas, which have a strong scientific focus, will be critical to designing and implementing any future national agri-environment scheme successfully over an enduring period.

Institutional learning is a slow and iterative process, with inevitable pockets of resistance to change. Governance and administrative arrangements need to be made more responsive and adaptive to implementing new policy innovations — with scientific underpinnings — if there is a genuine commitment to significant policy reform.

Policy reform will take a long time, and will be built step by step through innovation and experimentation. The Environmental Stewardship Program experience showed that, although governments demonstrate that they are sometimes willing to try new ways to protect biodiversity, their administrative arrangements are often so inflexible that they stifle the original innovative idea and approach. Innovative policy instruments need to be supported by more flexible and adaptive governance arrangements and an enduring commitment to credible scientific input. If this can be achieved, governments will be best placed to successfully implement market-based conservation initiatives.

Summary

Despite the challenges, we consider that the lessons learnt from the Environmental Stewardship Program can usefully inform future agri-environment schemes implemented by governments. Our key recommendations for any future public-funded and market-based program are:

1. Design and implement fit-for-purpose business processes and staff management strategies up front. Procurement plans that allow for ongoing provision of external services are needed, and land manager contracts should not be considered grants (and be regulated under grant guidelines). Rather, they should be commercial fee-for-service contracts to promote a business culture reflective of the service(s) the government is purchasing.

2. Focus on developing a more generic approach to maximising high-value biodiversity outcomes for as many priority investment targets (e.g. threatened species or ecological communities) as possible. A generic approach targeting habitat for multiple species/ communities will minimise the knowledge intensity of program design. That is, it will reduce the need for species-specific or community-specific conservation value metrics and management

plans. This approach could also easily accommodate the design principles of the Multiple Ecological Communities Project, including conservation management actions in the adjoining matrix at little additional cost (Whitten et al. 2011).

3. Any habitat-based approach should continue to develop greater landscape-scale connectivity between properties and across catchments (Zammit 2013b). This can be achieved by designing biodiversity markets that also support corridor development and continue to manage the agriculture matrix more sympathetically for conservation.

4. A greater emphasis needs to be given to the development of conservation management plans and appropriate performance assessment of their effectiveness. In particular, different grazing management strategies used by any new agri-environment scheme should be monitored and compared, to inform future programs.

5. A direct offers option can be implemented once sufficient market price information is available. Subject to considering other conservation priorities and the availability of funds, there is scope to offer additional interested land managers a fixed $/ha/year rate, and thereby improve efficiency and increase the area managed for biodiversity. A direct offer is a one-off offer of a contract by the government to a land manager with a stipulated price, duration, and management plan. The price is based on modelling price information from successful bids in previous tenders (see Binney and Zammit 2010 for a forest example). These land managers would hold an asset of quantified biodiversity value, as they would have participated in the initial aspects of a previous tender but either withdrew or were not successful. The subsequent fixed-price scheme would improve the return on investment for the program, which has high upfront costs because of the initial assessments, developing a suitable conservation value metric, and management plans.

Finally, a valuable outcome that the Commonwealth secured through this program (in addition to the hectares being managed) were the relationships forged with the contracted land managers and developed with the CSIRO and ANU. These relationships should be nurtured to foster further learning and trust (Gibbons et al. 2008). Effective conservation will come from mutual respect and common goals — the implementation phase of the program has provided the framework, but ongoing land manager support to 2025/26 is needed for enduring success. The more educational support provided to generate effective conservation outcomes, the more likely it is that land managers will believe in the benefits of a change in management practice. This will become critical for the maintenance of asset condition beyond 2025/26, although regulatory frameworks which prevent wilful degradation of protected assets will also play a role, as will conservation covenants for a number of properties.

Acknowledgements

The implementation of Environmental Stewardship Program from 2007 to 2012 was a success because of the dedication of the staff from the Environmental Stewardship section within the (now) Department of the Environment, as well as scientists from CSIRO and ANU, which informed different aspects of its design; and the Delivery Agents from Queensland, New South Wales, and South Australia, who worked tirelessly to implement the program within considerable time constraints.

We thank officers from the Department of the Environment for their time and willingness to discuss this program with the lead author in July 2014. However, we note that the department does not necessarily support or share the views expressed in this chapter.

References

Attwood, S.J. and E. Burns (2012) 'Managing Biodiversity in Agricultural Landscapes: Perspectives from a research-policy interface', *Land use intensification: Effects on agriculture, biodiversity and ecological processes* (eds D.B. Lindenmayer, S. Cunningham and A. Young), CSIRO Publishing, Melbourne, pp. 17–26.

Attwood, S.J., S.E. Park, M. Maron, et al. (2009) 'Declining birds in Australian agricultural landscapes may benefit from aspects of the European agri-environment model', *Biological Conservation* 142: 1981–91.

Binney, J. and C. Zammit (2010) 'The Tasmanian Forest Conservation Fund', *Paying for Biodiversity: Enhancing the cost effectiveness of payments for ecosystem services*, OECD, Paris.

Coggan, A., T.G. Measham, S. Whitten and D. Fleming (2013) *Socioeconomic monitoring for the environmental stewardship program*, report prepared by CSIRO Ecosystem Sciences for the Department of Sustainability, Environment, Water, Population and Communities, Canberra.

Ecker, S. and L.J. Thompson (2010) *Participation in the Environmental Stewardship program Box Gum Grassy Woodland Project: Key findings and implications*, ABARES, Canberra.

Gibbons, P., C. Zammit, K. Youngentob, et al. (2008) 'Some practical suggestions for improving engagement between researchers and policy-makers in natural resource management', *Ecological Management and Restoration* 9: 182–6.

Kay, G. M., D.R. Michael, M. Crane, et al. (2013) 'A list of reptiles and amphibians from Box Gum Grassy Woodlands in south-eastern Australia', *CheckList* 9(3), 476–81.

Kleijn, D. and W.J. Sutherland (2003) 'How effective are European agri-environment schemes in conserving and promoting biodiversity?' *Journal of Applied Ecology* 40(6): 947–69.

Lindenmayer, D.B., C. Zammit, S.J. Attwood, et al. (2012). 'A novel and cost-effective monitoring approach for outcomes in an Australian biodiversity conservation incentive program', *PLoS ONE* 7(12): e50872. DOI:10.1371/journal.pone.0050872.

Marsden Jacob Associates (2010) *Review of the Environmental Stewardship Program*, Department of Environment and Heritage, Canberra.

Morrison, T.H., C. McAlpine, J.R. Rhodes, A. Peterson and P. Schmidt (2010) 'Back to the future?: Planning for environmental outcomes and the new Caring for our Country program', *Australian Geographer* 41(4): 521–38.

Whitten, S.M., A. Langston, E.D. Doerr and V.A.J. Doerr (2011) *Real data testing of and improvements to the Multiple Ecological Communities Conservation Value Measure Tool*, final report for the Australian Government Department of Sustainability, Environment, Water, Population and Communities, CSIRO Ecosystem Sciences, Canberra.

Zammit, C.A. (2013a) 'Landowners and conservation markets: social benefits from two Australian government programs', *Land Use Policy* 31: 11–16.

Zammit, C.A. (2013b) 'Scaling up: The policy case for connectivity conservation and development of the National Wildlife Corridor Plan', *Linking Australia's Landscapes* (eds I. Pulsford, J. Fitzsimons, G. Wescott), CSIRO Publishing, Melbourne.

Zammit, C., S. Attwood and E. Burns (2010) 'Using markets for woodland conservation on private land: lessons from the policy-research interface', *Temperate Woodland Conservation and Management* (eds D.B. Lindenmayer, A.F. Bennett and R.J. Hobbs), CSIRO Publishing, Melbourne, pp. 297–307.

4

Do farmers love brolgas, seagrass and coral reefs? It depends on who's paying, how much, and for how long!

Geoff Park

Key lessons

- Productive farming can work hand in hand with environmental protection, especially when it is supported through understanding of farm-scale realities — such as a better understanding of the costs and risks associated with practice change — and better integration of biophysical, economic, and social knowledge.
- There needs to be better recognition of the extent to which improved private land management contributes to public good outcomes.
- The voluntary adoption of best management practices (BMPs) is unlikely to go far towards achievement of SMART environmental goals, as most required practices are simply not profitable or adoptable at the scale required.
- Successful outcomes will rely on the establishment of long-term financial incentives, in the form of stewardship payments to farmers, for the protection of important environmental assets.

- Evaluation of the ecological, social and economic implications of SMART targets is a critical step in understanding their attainability and desirability.

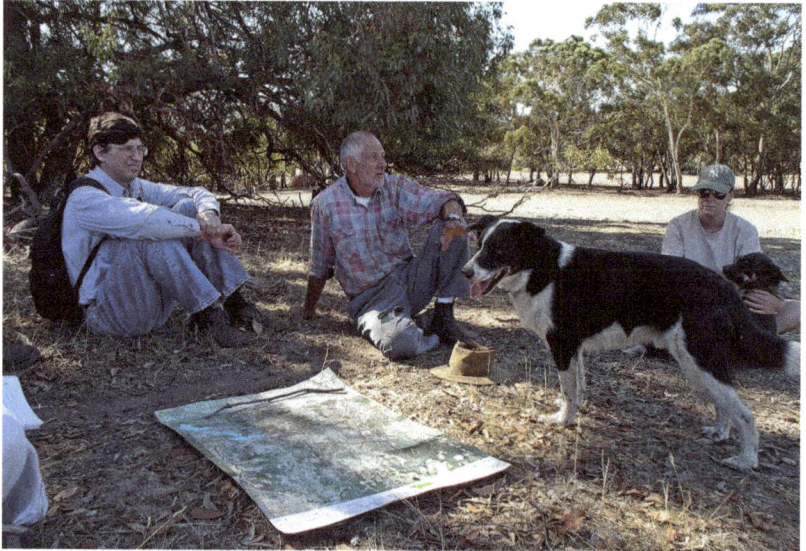

Figure 4.1: Farmers are better able to engage with environmental protection on their land where such work takes into account farm-scale realities.
Source: Photo by Geoff Park.

When it comes to farming and the environment, the rhetoric (some would say hyperbole) is full of claims about win–win outcomes. The truth, however, is less straightforward, and often not so convenient.

This chapter will reflect on this notion of win–win by examining the relationship between land management practices and the protection of significant environmental assets: natural assets of importance to the wider Australian community. Three case studies are provided, which are connected by a common thread: achieving significant and enduring environmental outcomes can only be contemplated if the public and private benefits (Pannell 2008) of land use and land management change are analysed and understood.

The first case study is set in the volcanic plains of central Victoria, the second in the catchment of Corner Inlet in Gippsland (south east Victoria), and the third lies northwards in the beef and sugar cane country that drains to the southern extremity of the Great Barrier Reef. Each case study spans dimensions of investment planning, strategic planning, and implementation. A feature of all three is the establishment of SMART targets that have been used to guide the design and evaluation of projects and plans. In all three case studies, an assessment was made using INFFER, the Investment Framework for Environmental Resources (Pannell et al. 2011, and see Box 4.1), to assess the cost-effectiveness of different scenarios, and to support the choice of appropriate actions and policy interventions. This was crucial, particularly in understanding the nature and scale of actions required by private landholders, the private costs associated with adopting these actions, and the technical feasibility and risks involved in generating the predicted environmental benefits.

What are SMART targets? There is some variation in how the letters of this acronym are defined. Our definition is widely used and simply asks if a target is Specific, Measurable, Attainable, Relevant, and Time-bound? The importance of SMART targets has been explicitly recognised by governments (Australian Government 2014).

Specific: The goal is described in a precise and unambiguous way.

Measurable: The goal definition is based on a variable which can be monitored and recorded reliably without unreasonable expense.

Attainable: A goal is more likely to be attainable when you plan your steps wisely and establish a time frame that allows you to carry out those steps. Thinking about attainable and realistic goals at the same time is useful.

Realistic: To be realistic, a goal must represent an objective toward which everyone is willing and able to work. A goal can be both high and realistic; the goal should represent substantial progress.

Time-bound: A particular date is provided by which time the goal will have been achieved. The time frame for the goal can be of any relevant duration. The time frame of achieving goals is related to the time for reasonable ecosystem response and costs.

The development and specification of SMART goals is crucial to understanding and analysing the costs and benefits of any proposed project (Doran 1981; McDonald and Roberts 2006). As a side note, I led a recent analysis that found that the use of SMART goals in planning by Australian regional natural resource management (NRM) bodies is not widespread and, in some cases, is getting worse (Park et al. 2013). Table 4.1 details the SMART goals that form the three case studies examined in this chapter.

Table 4.1: Summary of SMART goals respective case studies.

Case study	SMART goals
Moolort wetlands	By 2015 to: • Increase the extent of native vegetation surrounding the swamps on private land by 150 ha from 1,034 ha to 1,184 ha. • Improve the condition of native vegetation in and around the swamps on private land by 10 per cent (as measured by appropriate habitat assessment metric).
Corner Inlet	By 2032, reduce sediment and nutrient loads to: • Corner Inlet — nitrogen by 15 per cent, phosphorus by 15 per cent, and suspended sediment by 10 per cent. • Nooramunga — nitrogen by 10 per cent, phosphorus by 10 per cent, and suspended sediment by 15 per cent.
Great Barrier Reef and Great Sandy Straits — Burnett-Mary region	By 2033, to achieve: • 20 per cent overall reduction in anthropogenic suspended sediment load. • 20 per cent in anthropogenic loads of particulate nitrogen (PN) and particulate phosphorus (PP). • 50 per cent reduction in anthropogenic loads of dissolved inorganic nitrogen (DIN). • 50 per cent reduction in anthropogenic loads of dissolved inorganic phosphorus (DIP). • 60 per cent reductions of loads of PSII herbicides (hexazinone, ametryn, atrazine, diuron and tebuthiuron).

The Moolort Plains

The Moolort Plains consists of a chain of approximately 50 wetlands in the Loddon catchment, north central Victoria. The catchment of these wetlands is on the Victorian Volcanic Plain, and is Victoria's only national biodiversity hotspot (Australian Government 2015). The wetlands consist of freshwater meadows, shallow freshwater marshes, and deep freshwater marshes. No two wetlands in the network are the same; they vary in size (ranging from 1–2 ha to more than 200 ha in size), shape, and flora. These wetlands provide a range of breeding, roosting, and nesting sites for native fauna species, and are located in a largely cleared and fragmented landscape. A number of the wetlands are lignum dominated, providing excellent habitat for nesting waterbirds, including brolga.

The wetlands occur within the farm area, as islands of biodiversity, amongst intensive grazing and cropping systems. At the commencement of the project, few of the wetlands were fenced, with most grazed regularly or on an occasional basis. A number of the wetlands have been cropped in the past, and the possibility of future cropping is ever present. While many of the farmers who own these areas are driven, at least in part, by conservation values, the wetlands are seen as part of the farm production system, providing valuable fodder for animals, especially in times of drought, or as potential cropping land.

Figure 4.2: Frogmore Swamp is a wildlife refuge on the Moolort Plains.
Source: Photo by Geoff Park.

Preliminary discussions with the relevant landholders indicated that if projects demanding permanent changes in land use (e.g. cessation of grazing and cropping) and/or land management (e.g. fencing and changed grazing regimes) were to be successful, then landholders would need to be compensated for losses in production. This would require an appreciation of the economic production value of these areas before landholders would enter into agreements aimed at securing long-term conservation outcomes. On this basis, farm-scale economic assessments were undertaken to inform the design of a system of stewardship payments, tied to the establishment of conservation covenants.

As a result, some 400 hectares of significant wetland has been secured with permanent conservation covenants through the implementation of a project funded by the Australian Government. The level of funding (provided over a two-year period as works were implemented) through landholder stewardship agreements has been significant (on a per hectare basis), approaching current land values. Without this level of funding, it is doubtful that the project objectives would have been even partially achieved.

The remaining two case studies represent strategic planning projects for water quality, designed to inform future investment and implementation activities at sub-regional and regional scales.

Corner Inlet

Situated in south eastern Victoria, the Corner Inlet catchment is approximately 2,300km^2 in size and stretches along the South Gippsland coastline from Woodside to Wilsons Promontory. It is a highly productive area, supporting dairy, beef, and mixed grazing enterprises, and significant areas of production forestry. The region supports a significant Victorian commercial bay and inlet fishery, including 18 licensed commercial fishers. Corner Inlet itself supports outstanding environmental values that have been recognised through its listing as a wetland of international importance under the Ramsar Convention. The Ramsar area within Corner Inlet is the most southerly marine embayment and tidal mudflat system of mainland Australia, with extensive areas of seagrass supporting key ecological, economic, and social values.

Figure 4.3: Cows grazing in the Corner Inlet, near Wilsons Promontory, Gippsland, Victoria.
Source: Photo by Geoff Park.

Improvement of water quality in Corner Inlet requires a range of land management changes on farming lands, associated with the adoption of BMPs (e.g. maintaining ground cover and direct works, such as riparian fencing and revegetation) on agricultural land. Extension, positive incentives, and regulation compliance activities have all been used in Corner Inlet.

For activities on agricultural land aimed at improving water quality, most programs have been focused on incentive and extension activities to influence the implementation of actions and the adoption of BMPs. Some of these programs operate in tandem; for example, extension activities often identify on-ground works, such as waterway fencing, which are then implemented through direct grant programs. Likewise, incentive delivery is generally coupled with extension information for landholders outlining appropriate maintenance activities. Historically, these programs have been successful in engaging landholders in the implementation of actions and the adoption of BMPs, and have been delivered in a collaborative way across agencies.

Initially, in the development of the Corner Inlet Water Quality Improvement Plan (WQIP) (Dickson et al. 2013), a set of aspirational targets were analysed. These targets, requiring significant reductions in the load of phosphorus (30 per cent), nitrogen (30 per cent) and sediment (10 per cent), were deemed to be crucial to protect the ecological values of the area.

However, bio-economic modelling results indicated (Dickson et al. 2013) that while these targets were theoretically feasible, they would require large-scale changes in land use (e.g. conversion of grazing land to native vegetation), together with long-term stewardship payments to support landholder adoption of BMPs. Not only would the cost of these changes have been very large (in the order of $35 million annually) they would have invoked unacceptable socio-political risks, such as those associated with conversion of marginal farm land into permanent conservation areas. Consequently, a set of more modest SMART targets were used (see Table 4.1) to inform the future implementation of works to protect water quality in Corner Inlet. These targets could be reached without land use change, but still require ongoing payments for BMP adoption, together with significant waterway and erosion works.

The degree to which these modest targets will support a significant ecological response (e.g. an increase in the extent and condition of seagrass) remains a matter of conjecture, as the science regarding ecologically relevant targets for Corner Inlet is equivocal (Dickson et al. 2013).

What is clear is that the level of funding available through current programs (approximately $1 million annually) is not sufficient to achieve the required reduction in nutrient and sediment entering Corner Inlet, and that a scaled-up program (in the order of $4 million annually) is required to meet the modest, but agreed, implementation targets that form the basis of the WQIP. This is not surprising given the findings of analyses for similar environments, such as the Gippsland Lakes, where it has been estimated that a 40 per cent reduction in phosphorus entering the system would require around $1 billion over 25 years (Roberts et al. 2012), well beyond the current levels of funding available for catchment scale works.

The Great Barrier Reef and Great Sandy Straits

The Great Barrier Reef and Great Sandy Straits areas are located adjacent to the Burnett-Mary NRM region, in southern Queensland. The Burnett-Mary region contains a diverse range of riverine, coastal, and marine habitats. Included in the region is the southernmost portion of the World Heritage-listed Great Barrier Reef Marine Park, and the Ramsar-listed Great Sandy Strait, which host biodiversity values that are globally important, including important populations of threatened dugong.

The Burnett-Mary region has an approximate catchment area of 56,000 km^2 and is approximately 12 per cent of the total Great Barrier Reef catchment area (423,122 km^2). The health of the marine environment is affected by a range of short-term and chronic longer term threats. The major pressures and threats include terrestrial pollutants (sediment, nutrients and pesticides), coastal development, shipping (and boating), fishing/netting, and climate change. Collectively, terrestrial runoff (and resulting pollutant loads and poor water quality) is considered to have the most effect on coastal and marine assets in the region.

The experience in the Burnett-Mary region has a number of parallels with Corner Inlet. In the development of the Burnett-Mary WQIP, an initial set of ecologically relevant targets were set for nutrients, sediments, and pesticides. Again, the cost and feasibility of achieving these targets was assessed using a bio-economic model, with the results suggesting an annual investment of $16 million would be required to support BMP adoption in the sugar cane and grazing industries. While not requiring land use change, as was the case for Corner Inlet, this result was deemed to be unrealistic, given current and likely future levels of funding. Consequently, a set of revised SMART targets (see Table 4.1), based on the Reef Plan Water Quality Protection Plan (Secretariat for Reef Water Quality Protection 2013), were used to underpin the WQIP. Reaching these targets is still challenging, with an annual investment of $5 million, and would require a targeted approach to BMP support, largely involving the sugar cane industry rather than grazing, and with areas of specific geographic focus.

As with Corner Inlet, the extent to which the agreed targets would protect the key ecological values of the Great Barrier Reef and Great Sandy Straits requires further research.

Box 4.1: INFFER — Investment Framework for Environmental Resources

INFFER™ is a tool for developing and prioritising projects to address environmental issues such as reduced water quality, biodiversity, environmental pests, and land degradation. It is designed to help environmental managers achieve the most valuable environmental outcomes with the available resources.

It consists of a seven-step process, which begins by identifying significant assets and works through project development, project assessment and selection, implementation, and, finally, monitoring, evaluation, and adaptive management. (For the logic behind these steps, see Chapter 22, in which David Pannell reflects on the how the performance of agri-environment programs can be improved.)

The Project Assessment Form, completed in step three, is the key component of the process. This is where users record information about the asset, the threats compromising it, the goals that the project will achieve, and the actions needed to achieve those goals. Judgements about the likelihood of success in terms of technical feasibility and community and government support are also made and recorded here, and the proposed project budget is specified. The core information is used to calculate a benefit–cost ratio that indicates the project's value for money.

Step 1: Identifying significant assets

A list of significant natural assets that are candidates for investment is prepared. These assets can be drawn from existing documents or lists, from community workshops, from relevant experts, or from analytical processes, such as systematic conservation planning. At the regional level, the list may include 100 to 300 significant assets.

Step 2: Filtering significant assets

Using a simplified set of criteria, the list of significant assets is filtered down to ~20–40 assets. Our suggested approach is to identify assets of high significance, with high current or predicted future damage. The filtered list is further reduced (to ~10–20) using the five questions on the INFFER pre-assessment checklist. Assets may be culled at this point because they are not spatially explicit, because a specific, measurable, time-bound goal cannot be formulated, or because an initial assessment indicates that the project would not be cost-effective.

Step 3: The Project Assessment Form

Using the INFFER Project Assessment Form, develop an internally consistent project for each asset on the reduced list. This process draws together readily available information, consisting of desktop review of publications and reports, and consultation with the community and relevant experts. Information required at step three includes asset significance, threats, project goal, works and actions, time lags, effectiveness of works, private adoption of actions, delivery mechanisms, and costs. Using this information, apply the Public: Private Benefits Framework to help select policy mechanisms, and calculate a benefit–cost ratio to be used in project ranking. The output from step three is a Project Assessment Report, which includes the benefit–cost ratio, risk factors (practice change, technical feasibility, socio-politics, long-term funding), spin-offs, quality of information, and key information gaps.

Step 4: Selection of priority projects

Select a short list of priority assets/projects based on the information in the Project Assessment Report and other relevant considerations.

Step 5: Investment plans or funding proposals

Develop investment plans or proposals for external funding (depending on whether INFFER is being used to allocate an internal budget or to develop and assess projects for external funding).

Step 6: Implementing funded projects

Implement those projects that receive funding. In many cases, the first stage of a project should consist of a detailed feasibility investment, involving targeted collection of additional information to strengthen the assessment done in step three.

Step 7: Monitoring, evaluation, and adaptive management

Monitor, evaluate, and adaptively manage projects. After feasibility assessment, and at regular intervals thereafter (every two years, for example), the data in the original Project Assessment Form for each funded asset/project should be updated to reflect lessons learned, progress towards outcomes, and any new data or analysis that has become available. At this point, managers should consider whether the original design of the project is still suitable, and whether the project should remain a priority.

More information: www.inffer.com.au

Conclusion

These case studies highlight the important role that farmers play in protecting significant environmental assets at a range of scales. While these assets occur in the private land estate and on public land, it is the actions of farmers that have a principal impact on the achievement of public good outcomes. Achievement of these outcomes requires significant, long-term changes in land use and management, which come at considerable financial and social cost to farmers, and will not be adopted without the provision of adequate financial incentives in the form of stewardship payments.

In all three case studies, an assessment was made using INFFER, the Investment Framework for Environmental Resources (Pannell et al. 2011, and see Box 4.1), to assess the cost-effectiveness of different scenarios, and to support the choice of appropriate actions and policy interventions. Unsurprisingly, the detailed analysis supported by INFFER revealed that the actual cost of the described projects was far greater than originally anticipated, perhaps by an order of magnitude. The specification and analysis of SMART targets is crucial to understanding the scale of payments required to meet ecologically and socially meaningful results.

In both water quality projects examined, not only are the direct upfront costs considerably greater than could be accommodated through current state and Commonwealth environment programs, but there are also significant ongoing costs associated with maintaining benefits. The payment of ongoing stewardship costs, especially when linked to BMPs, is likely to be challenging and require improved scientific and technical understanding of the effectiveness of practices and a major shift in Australian NRM policy.

References

Australian Government (2014) *Monitoring, evaluation, reporting and improvement tool (MERIT)*, Commonwealth of Australia, Canberra. Available at: www.nrm.gov.au/funding/merit/index.html.

Australian Government (2015) *Australia's 15 National Biodiversity Hotspots,* Commonwealth of Australia, Canberra. Available at: www.environment.gov.au/biodiversity/conservation/hotspots/national-biodiversity-hotspots.

Beverly, C., A. Roberts, K. Stott, O. Vigiak and G. Doole (2013) 'Optimising economic and environmental outcomes: Water quality challenges in Corner Inlet Victoria', *Proceedings of MODSIM 2013*, Adelaide. Available at: www.mssanz.org.au/modsim2013.

Dickson, M., G. Park, and A. Roberts (2013) *Corner Inlet Water Quality Improvement Plan*, West Gippsland Catchment Management Authority, Traralgon, Victoria.

Doran, G.T. (1981) 'There's a S.M.A.R.T. way to write management's goals and objectives', *Management Review* 70(11): 35–6.

McDonald, G. and B. Roberts (2006) 'SMART water quality targets for Great Barrier Reef catchments', *Australasian Journal of Environmental Management* 13: 95–107.

Pannell, D.J. (2008) 'Public benefits, private benefits, and policy intervention for land-use change for environmental benefits', *Land Economics* 84(2): 225–40.

Pannell, D.J., A.M. Roberts, G. Park, et al. (2011) 'Integrated assessment of public investment in land-use change to protect environmental assets in Australia', *Land Use Policy* 29: 377–87.

Park, G., A. Roberts, J. Alexander and D. Pannell (2013) 'The quality of resource condition targets in regional natural resource management in Australia', *Australasian Journal of Environmental Management* 20(4): 285–301. Available at: dx.doi.org/10.1080/144 86563.2013.764591.

Park, G., C. Beverly, A. Roberts and M. Dickson (2014) *Bioeconomic modelling scenarios and results report: Milestone 3 — Burnett Mary Water Quality Improvement Plan*. Available at: www.bmrg. org.au/files/4314/3700/8083/BE_Model_Scenarios_Final_report-28Mayv6.pdf.

Roberts, A.M., D.J. Pannell, G. Doole and O. Vigiak (2012) 'Agricultural land management strategies to reduce phosphorus loads in the Gippsland Lakes, Australia', *Agricultural Systems* 106(1): 11–22.

Secretariat Reef Water Quality Plan Protection (2013) *Reef Water Quality Protection Plan*, Queensland Government. Available at: www.reefplan.qld.gov.au/resources/assets/reef-plan-2013.pdf.

5

The vital role of environmental NGOs: Trusted brokers in complex markets

David Freudenberger

Key lessons

- Complex markets need brokers; eNGOs have performed this role well.
- Innovation is critical, and this requires organisations willing to fail. eNGOs are important for innovation.
- Engaging farmers is important, but sometimes it is not enough.
- Making a profit is OK.
- eNGOs provide a voice for the voiceless.

Environmental non-government organisations (eNGOs) have been actively involved in agri-environment schemes since their inception in Australia. These include a range of groups — large and small — operating over a range of scales. Their role and value is sometimes overlooked in discussions on agri-environment schemes, yet their contributions are profound and often critical to the success of the scheme. In this chapter, I outline why eNGOs are so important to agri-environment schemes and list five key lessons that should always be kept in mind by the designers of future schemes.

Figure 5.1: Greening Australia conducts a WOPR field day. Farmers and NRM officers are being shown over one of the first WOPR sites.

Source: Photo by David Salt.

Who are we talking about when we speak of eNGOs? At the national and international end of the scale we have organisations such as Greening Australia, Landcare Australia, Conservation Volunteers Australia, WetlandCare Australia, Birdlife Australia, World Wildlife Fund (WWF), Australian Conservation Foundation, Bush Heritage Australia, Australian Wildlife Conservancy, The Wilderness Society, and The Nature Conservancy. While all of them have an environmental dimension, they are quite different characters with different areas of focus, different people, skill sets, and modus operandi.

In addition to these large eNGOs, many hundreds of state-based, regional, and local non-government organisations have also been deeply involved with Australian agri-environment programs for over two decades. State-based organisations include trust-for-nature organisations established under state legislation in New South Wales, Victoria, and Queensland. Regional organisations include quasi-government natural resource management (NRM) organisations, such as catchment management authorities (CMAs) in Victoria. At the local scale, Landcare groups and networks, many of them incorporated, have played a pivotal role in shaping and implementing government agri-environment schemes.

Over the past two decades, I have gained insights from being involved with many of these organisations operating at all scales (international to local), although I am biased by having been Chief Scientist with Greening Australia for five years (2007–2012) and having collaborated with Greening Australia as a research scientist with CSIRO. I currently collaborate with Greening Australia and Bush Heritage Australia as an academic at ANU. So here is my shortlist of insights from 20 years of being an insider and outsider to many of the eNGOs involved in Australian agri-environment schemes.

Complex markets need brokers

Agri-environment schemes are a market with essentially one buyer (the government) and a great diversity of many sellers (farmers). Australian governments essentially purchase public good environmental outcomes from many thousands of private agricultural enterprises that manage over 50 per cent of the Australian continent. Government purchases environmental outcomes primarily through direct or indirect grants to farmers. These grants generally subsidise the cost of inputs, such as planning, fencing, pest control, and revegetation (see Chapter 2 by Graham Fifield), as well as some organisational overheads. Like any buyer, the government has choices: direct one-to-one purchases of public goods from farmers, or one-to-many purchases through a diversity of brokers and arrangements. These two broad options are no different to an individual buying company shares on the stock market directly, or through the expertise (and additional cost) of a stockbroker.

A few national eNGOs, such as Greening Australia and WWF, pioneered the role of broker. Starting in the mid-1980s, they received a few large contracts from the Australian Government and delivered devolved grants to farmers through simple one-stop shop programs. These eNGOs handled the reporting obligations to the federal government; the farmers did the work on-ground, with planning advice and implementation assistance from the eNGOs.

This brokering or facilitation role was particularly important in the early agri-environment schemes. At the time, farmers were very reluctant sellers to a distant and seemingly suspicious national government buyer, with little capacity to get staff out into the paddock and into the kitchen for a cup of tea. At one point in the mid-2000s, Greening Australia had more than 350 staff and over 30 regional offices brokering thousands of public-good environmental purchases, facilitated through many paddock walks and cups of tea. Similarly, WWF secured funding and had regional staff who facilitated the national Threatened Species Network, which help create awareness of threatened species on private land and brokered many small grants to improve their conservation.

This brokering role has generally been taken over by, at one point, 57 NRM organisations, mostly based in regional Australia and set up by state legislation and funded or co-funded by the federal government (Curtis et al. 2014). However, particularly in New South Wales, these NRM organisations have become another branch of government, as they are responsible for, among other things, the administration of land clearing legislation. It can be argued that eNGOs are rarely fully independent of government, since so many receive government funding. But they have been shaping and delivering many agri-environment schemes on behalf of Australian governments over the past 30-plus years.

The advantage of engaging a broker is the ability to build lasting relationships to help navigate the complexities and risks of entering and persisting in any market. Many eNGOs have persisted through decades of agri-environment schemes that often don't last for more than one election cycle. Continuity and organisational identity is a strength of many eNGOs. The environmental market is particularly complex, with high overheads (e.g. reporting to the government funder). Many eNGOs provide this brokerage service to government (the risk-averse buyer) and farmers (the risk-averse sellers). Without such brokers, the public good market for environmental services and outcomes would likely have been slower to develop in Australia.

Innovation: Willingness to fail

Environmental NGOs involved in Australian agri-environment schemes have a demonstrated record of risk-taking leading to outstanding innovations. They are small enough to take on risky research and development, but large enough to survive small failures. I assert that many technical innovations in securing environmental outcomes originated from or were nurtured by a diversity of local, regional, national, and international eNGOs. Here are a few notable examples:

- *Direct seeding*: Direct seeding technology that has successfully reduced the cost and increased the diversity of native vegetation restoration was developed by local farmer innovators and rural engineers, facilitated and marketed by the likes of Greening Australia, with support from corporate partners and government agencies.

- *Fencing*: Planting fences around woodland remnants (e.g. fencing incentive schemes) rather than just planting trees was rigorously promoted to the Australian Government during the 1990s by local Landcare groups and Greening Australia, who had the capacity to lobby government in Canberra. Only later did research organisations test the effectiveness of these fencing programs to protect and enhance remnants of native vegetation (Spooner et al. 2002; Briggs et al. 2008).

- *Seed production*: Greening Australia made a notable contribution to the development of commercial-scale seed banks and on-farm seed production areas.

- *Restoration and rehabilitation*: Greening Australia facilitated the technological innovations needed to restore, at scale, species-rich temperate grasslands (Gibson-Roy et al. 2010). The Whole of Paddock Rehabilitation (WOPR) program was invented by a farmer, but was marketed and rolled-out by Greening Australia (see Chapter 15 by Dean Ansell and Chapter 2 by Graham Fifield).

- *Floodplain management*: The then Murray Wetlands Working Group pioneered small-scale floodplain restoration, including the management of two environmental water licenses, recognised by the award of the prestigious National Thiess Riverprize in 2007 (www.murraydarlingwetlands.com.au).

- *Conservation planning*: The Nature Conservancy played a pivotal role in introducing conservation action planning to Australia (now termed 'Open Standards for the Practice of Conservation' — cmp-openstandards.org). This is a participatory planning process that is particularly effective at engaging a diversity of stakeholders in setting conservation objectives.

- *Corridors*: Local, regional, and international NGOs have been particularly influential in developing large-scale corridor initiatives, such as Gondwana Link (www.gondwanalink.org), Habitat 141⁰ (www.habitat141.org.au), and the Bunya Biolink (Freudenberger et al. 2013). Only much later did the federal government develop the National Wildlife Corridor Plan (www.environment.gov.au/topics/biodiversity/biodiversity-conservation/wildlife-corridors).

As environmental brokers, eNGOs are in the position to recognise and facilitate innovation addressing local, regional, and national environmental challenges.

Figure 5.2: Direct seeding of trees has successfully reduced the cost and increased the diversity of native vegetation restoration.

Source: Photo by Greening Australia.

Engaging farmers is sometimes not enough

In many agricultural landscapes, significant conservation gains cannot be achieved just by fencing off a few on-farm remnants and planting shelter belts. In places, there is a need to buy the farm and restore the lot for public good conservation outcomes.

Environmental NGOs, including the Tasmanian Land Conservancy, Bush Heritage, and Australian Wildlife Conservancy (and Greening Australia to a lesser extent), are doing just that — purchasing farms and pastoral properties in strategic locations and engaging in long-term restoration and conservation. To date, Bush Heritage Australia works across more than 4.8 million ha, including their own reserves and land owned by their partners. The Australian Wildlife Conservancy owns and manages 23 properties covering more than 3 million ha. The trusts for nature in Queensland, New South Wales, and Victoria also facilitate private investment in purchasing farmland and converting tenure to conservation in perpetuity.

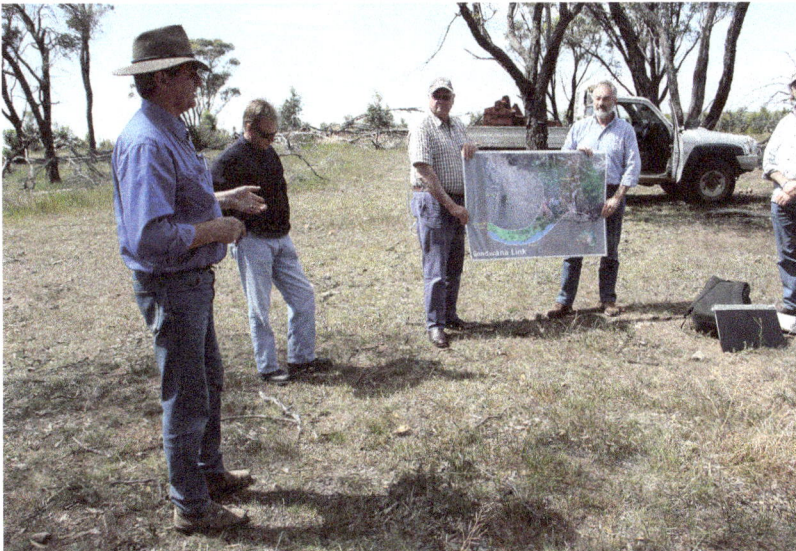

Figure 5.3: eNGOs have purchased whole properties in strategic locations in southern Western Australia. These have formed the backbone of the biolink called Gondwana Link.
Source: Photo by David Salt.

The entrepreneurial nature of many eNGOs allow them to be free to operate at a variety of scales and locations that would be difficult for government. These agricultural land purchases and environmental management changes by eNGOs contribute to the public estate of nature reserves, but have far greater flexibility in funding, purchasing, and management than is possible for government. In general, purchases by eNGOs appear to create much less local angst than government agencies purchasing farmland on the open market. The important role of eNGOs in contributing to the national conservation estate was recognised by the Australian Government's National Reserve System Program, which co-invested in NGO conservation land purchases. For example, the National Reserve System Program has contributed more than $8 million to Bush Heritage Australia since 1999 to assist in the purchase of 11 protected areas covering 890,000 ha. This program has also invested more than $430,000 with the Tasmanian Land Conservancy and $6 million to the Australian Wildlife Conservancy for land purchases (www.environment.gov.au/land/nrs/getting-involved/conservation-organisations).

Making a profit is OK

Government and quasi-government environmental organisations are constrained by the political vagaries of annual budget allocations and zero-sum accounting (e.g. annual funding must be fully acquitted within the financial year). Environmental NGOs, as registered charities or not-for-profit businesses, can carry over funds from one year to the next. They can act as businesses and make a surplus (profit) that must be re-invested within the organisation (rather than paid out to private shareholders). NGOs can (and do) build financial reserves to support long-term strategic programs, as well as using reserves to maintain staff continuity and corporate memory (Maier et al. 2014).

Such flexibility and governance structures (e.g. boards) continue to attract philanthropic investment from individuals and foundations. Philanthropy rarely donates funds to government agencies; rather, donations to eNGOs are effective in leveraging government co-investment (see www.abc.net.au/radionational/programs/bush telegraph/private-park-funding-slows/5888378). The history of Australian agri-environment schemes includes long-term involvement of philanthropy, from the early days of Potter Foundation Farmland Plan (www.ianpotter.org.au/potter-farmland-plan) to the many thousands of individual small-scale donations to eNGOs, such as the Nature Conservancy, Bush Heritage Australia, and the Australian Wildlife Conservancy actively conserving biodiversity in agricultural landscapes across Australia.

A voice for the voiceless

Environmental NGOs continue to have a significant role in lobbying governments to direct funding to the environment, including agri-environment schemes. The original National Landcare Program was initiated by a proposal arising from collaboration between the Australian Conservation Foundation and the National Farmers Federation. WWF gave a voice to Australia's highly depleted temperate grasslands and was successful in securing government funding for grassland conservation, assisted by devolved grants to farmers. NGOs have had a significant role in including farmland in Ramsar-listed wetlands, such as the Gwydir and Macquarie Marshes in central NSW. Environmental NGOs have been influential in highlighting the global significance of the Great Western Woodlands of south west Western Australia (Watson et al. 2008) and Australia's tropical savannas (Woinarski et al. 2007). Both these regions are dominated by agricultural (pastoral) uses.

As brokers between governments and private landholders, many eNGOs also provide a collective voice to progressive (green) farmers who may not be well represented by their often highly politicised state and national agricultural organisations, who are more focused on agricultural productivity issues than public good nature conservation. Many eNGOs have the time and talent to provide this voice through a myriad of conversations in capital cities where political power and funding reside.

Conclusion: Still a role

Based on my personal involvement with agri-environment schemes and eNGOs over the past 20 years, I assert that a remarkable diversity of eNGOs have shaped and continue to shape agri-environment schemes. These eNGOs have likewise been shaped by these schemes, largely funded by governments. These schemes would have been far less innovative, flexible, and responsive without this history of NGO involvement. eNGOs have a critical role in continuing to deliver innovation, efficiencies, and effectiveness in supporting significant public good outcomes from government investment in conservation on-farm.

Acknowledgements

The insights and opinions provided in this essay are those of the author, but are based on many conversations over the past two decades with NRM leaders, particularly those working in eNGOs, including Martin Driver, Ian Davidson, Owen Whitaker, Jaimie Pittock, David Williams, Carl Binning, David Butcher, Graham Fifield, Jonathon Duddles, Toby Jones, Robert Lambeck, Dave Carr, Keith Bradby, and Sue Streatfield. Comments from two anonymous peer reviewers have helped to sharpen and refine this essay. Thanks to you all.

References

Briggs, S.V., N.M. Taws, J.A. Seddon and B. Vanzella (2008) 'Condition of fenced and unfenced remnant vegetation in inland catchments in south-eastern Australia', *Australian Journal of Botany* 56: 590–9.

Curtis, A., H. Ross, G.R. Marshall, C. Baldwin, et al. (2014) 'The great experiment with devolved NRM governance: Lessons from community engagement in Australia and New Zealand since the 1980s', *Australasian Journal of Environmental Management* 21: 175–99. DOI:10.1080/14486563.2014.935747.

Freudenberger, D., L. Grigg and R. Reeger (2013) 'The Bunya BioLink: An application of Greening Australia's strategic approach to large scale conservation', *Linking Australia's Landscapes* (eds J. Fitzsimmons, I. Pulsford and G. Wescott), CSIRO Publishing, Collingwood, pp. 153–62.

Gibson-Roy, P., G. Moore, J. Delpratt and J. Gardner (2010) 'Expanding horizons for herbaceous ecosystem restoration: The Grassy Groundcover Restoration Project', *Ecological Management and Restoration* 11: 176–86.

Maier, F., M. Meyer and M. Steinbereithner (2014) 'Nonprofit organizations becoming business-like: A systematic review', *Nonprofit and Voluntary Sector Quarterly* December. DOI:10.1177/0899764014561796.

Spooner, P.G., I.D. Lunt and W. Robinson (2002) 'Is fencing enough?: The short-term effects of stock exclusion in remnant grassy woodlands in southern NSW', *Ecological Management and Restoration* 2: 117–26.

Watson, A., S. Judd, J. Watson, A. Lam and D. Mackenzie (2008) *The Extraordinary Nature of the Great Western Woodlands,* Wilderness Society of WA Inc., Perth. Available at: www.wilderness.org.au/sites/default/files/media/the-great-western-woodlands-report.pdf.

Woinarski, J., B. Mackey, H. Nix and B. Traill (2007) *The nature of northern Australia: Natural values, ecological processes and future prospects,* ANU E Press, Canberra. Available at: press.anu.edu.au/nature_na/pdf/whole_book.pdf.

6

Agricultural land use policy in the European Union: A brief history and lessons learnt

Rob Fraser

Key lessons

- Unforeseen consequences: the origins of the European Union's land use policy was in setting aside agricultural land as an instrument of production control. It was not aimed at generating environmental benefits, and yet it ended up doing this as well.

- Carrots versus sticks: the European Union has felt (increasingly) socially empowered to use policy sticks on its farmers to deliver improved environmental goods and services from the agricultural landscape, whereas in the US and Australia policymakers have felt obliged to offer mainly carrots to farmers to do so.

- Environmental benefits versus foregone agricultural income: the UK is in the process of introducing a new Countryside Stewardship Scheme whereby applications to participate will be scored according to their fit with empirically validated attributes of environmental benefit. As a consequence, the cost-effectiveness of environmental stewardship in delivering benefits to society will be improved.

Figure 6.1: Cattle in a UK field ringed by stone walls.
Source: Photo by Colin Virden.

This chapter provides both a brief history of agricultural land use policy as a component of the Common Agricultural Policy (CAP) of the European Union (EU) and a discussion of some lessons learnt in this context.

Note that the term 'agricultural land use policy' refers to the component of the CAP which is now widely referred to as agri-environmental policy, because it relates (in policy terms) to the impact of food production on the environment **within the specific context of the agricultural landscape**. It focuses primarily on the impacts associated with habitats, species (flora and fauna), and water and soil quality rather that effects on other aspects of the environment such as the air or ocean — it is not concerned with greenhouse gas emissions, for example.

A brief history

The CAP's agricultural land use policy has a comparatively short history, having been introduced in the form of voluntary set-aside as a production control mechanism in the CAP's 'crisis years' of the late 1980s, when the cost of funding ongoing production surpluses of the EU's major farm commodities threatened to derail the CAP's budget. In this context, the most comprehensive and up-to-date account of the history of the CAP is provided by the European Commission on its website: ec.europa/agriculture/cap-history/index_en.htm.

According to the European Commission, the CAP is characterised as comprising three main stages:

i.   from 1957 — productivity

ii. from the 1992 CAP reform — competitiveness

iii. from the Agenda 2000 CAP reform — sustainability

In this account, the CAP's agricultural land use policy is formally introduced as a compliance land set-aside program within the package of the 1992 CAP reforms. Its aim was to improve competitiveness in EU agriculture. (In this program, farmers were required to take a stipulated proportion of their arable land out of production in order to receive compensation payments for reductions in previous (guaranteed) levels of price support.) This characterisation overlooks the earlier introduction of voluntary set-aside (in 1988), but in so doing emphasises the new (in 1992) requirement of set-aside in order to receive compensatory payments for reductions in price support, which were central to the 1992 CAP reform package.

Moreover, the sustainability characterisation represents the formal creation of 'Pillar 2' as part of the CAP, and the explicit recognition of farmers as environmental stewards, who became (in 2003) not just required to keep their land in good agricultural and environmental condition in order to receive direct payments (previously called compensatory payments), but were also offered further financial incentives to protect and enhance their provision of environmental goods and services on their land. ('Pillar 1' is the term used to refer to EU budget funding for direct payments; 'Pillar 2' is the term used to refer to EU budget funding for environmental stewardship.)

Looking ahead, as the CAP moves towards 2020 we can already see proposals to shift more of the CAP's budget from Pillar 1 to Pillar 2, thereby further strengthening the role of the CAP's agricultural land use policy as one of supporting environmental sustainability, with farmers as central providers of environmental benefits from this support.

The goal of productivity

If we go back to the 1980s, to what the European Commission calls the CAP's 'crisis years', with the benefit of hindsight we can clearly see the beginnings of the agricultural land use policy process by which farmers have now come to be seen as environmental stewards, rather than destroyers of the environment, as they were at the time.

Recalling that the first stage of the CAP is characterised by the European Commission as 'productivity', we are reminded that one of the principle objectives of creating the European Economic Community in 1957 was to deliver food security. As a consequence, farmers were encouraged to increase their production with a range of market intervention measures designed to provide price support (e.g. tariffs, intervention purchasing). In response to this price support, farmers did take steps to increase their production, both by the intensification (increasing yields on a given area of land) and extensification (increasing the area under agricultural production) of land use.

This productivity stage continued successfully through the 1960s and 1970s, at which point ongoing production surpluses began to become apparent, initially in the dairy sector but by the early 1980s extending across the range of the EU's major farm commodities.

These production surpluses can now be seen as the cause of two separate concerns that developed during the 1980s. The first was the policy concern relating to the CAP's budget, which was required to fund the (supported) prices of surplus farm commodities. The second was a social concern, relating to the perceived negative effect of the intensification and extensification of agricultural land use on the environment.

The first concern led to the realisation that steps needed to be taken to control farm production within the EU. How might this be achieved? Across the Atlantic, the US had developed a land-diversion policy as a production control mechanism in the 1985 Farm Bill (see Ervin 1988). Following this lead, the EU took its first step into the domain of an agricultural land use policy by introducing voluntary set-aside in 1988. This embryonic policy offered a carrot to farmers in the form of set-aside payments to take a proportion of their land out of production. However, given the level of these payments compared with the foregone production income (with supported prices) from set-aside land, the incentive for farmers to engage with the policy was very weak, and the uptake was therefore very low.

Environmental concerns

Meanwhile, developing alongside this policy concern was the social concern about environmental degradation caused by the intensification and extensification of agricultural land use. In relation to intensification, supported prices gave farmers the incentive to increase yields with the addition of fertiliser, leading specifically to the problem of nitrate leaching affecting groundwater. This was of particular concern in areas where groundwater was used to provide potable water for domestic consumption. In addition, this price support encouraged farmers to maximise the area of their land under production, thereby leading to the destruction of habitats (e.g. hedgerows, native woodland).

This social concern gave rise to calls in the academic literature for conservation set-aside to be introduced — to deliver a policy win–win by encouraging farmers to take land out of production which would also deliver environmental benefits (see, for example, Gasson and Potter 1988). While there is no doubt that the formal introduction of voluntary set-aside as an agricultural land use policy in 1988, and its modification in the 1992 CAP reform to compliance set-aside, was driven primarily by the EU's ongoing production surpluses and the associated crisis in the CAP's budget, this social concern in the 1980s was undoubtedly a precursor to the subsequent policy process in the late 1990s. This led to the introduction of the environmental 'sustainability' stage of the CAP with the Agenda 2000 reforms.

While this social concern about the negative impact of farming on the environment was developing further into the 1990s, so was the policy awareness that the CAP's voluntary set-aside scheme was not delivering sufficient production control. In this context, the EC's website detailing the history of the CAP (see above) contains an excellent package of information about the development of the 1992 CAP reform (known also as the MacSharry reform).

Specifically, it was acknowledged that price support was at the centre of the production surplus problem, and so this needed to be reduced, encouraging the de-intensification of agricultural land use. In addition, the evolution of the CAP's agricultural land use policy from voluntary to compliance set-aside was intended to encourage the de-extensification of agricultural land use. The only risk to compliance set-aside was that farmers would choose to forego their compensatory payment for reduced price support in order to keep all their land in production. As it turned out, this risk proved to be extremely low.

As a consequence, the implementation of the 1992 CAP reform saw the amelioration of the EU's production surplus problem. And while compliance set-aside played its role in this process, academic analysis of the land-use response of farmers to their set-aside requirement also revealed the policy win–win anticipated by the movement for conservation set-aside in the 1980s. This situation applied particularly to what was called 'non-rotational set-aside', whereby land was set-aside for at least five years. This was always the set-aside option widely preferred by farmers.

Striving for sustainability

This reduced concern about the CAP budget, combined with the growing social awareness of the environmental impacts of agricultural land use, led to increased support for further CAP reform to raise the profile of environmental considerations in its operation — hence the 'sustainability' stage of the CAP introduced with the Agenda 2000 reform.

The creation of Pillar 2 was central to the Agenda 2000 CAP reform. Pillar 2 provided explicit financial support for the 'integration of environmental concerns into agricultural policy' (see the Agenda 2000 reform page of the European Commission's *History of the CAP* website). Although set-aside was retained as an agricultural land use policy within this reform, increasingly farmers were encouraged to see their set-aside land in terms of the policy win–win — production control plus environmental benefit.

Moreover, as previously noted, this policy impetus towards environmental stewardship by farmers was maintained with the 2003 CAP reform, which both de-coupled direct payments from production and introduced cross-compliance, whereby farmers were required to keep their land in good agricultural and environmental condition in order to receive direct payments. In addition, there was a further shift of CAP financial support from Pillar 1 to Pillar 2 — known as 'modulation', resulting in the development of voluntary environmental stewardship schemes such as the UK's Higher Level Stewardship Scheme.

So successful was the refocusing of the CAP's agricultural land use policy towards environmental sustainability that the decision was taken in 2008 to abolish set-aside. By this time, production surpluses were a thing of the past, and social support for farmers to be incentivised to protect and enhance environmental goods and services had become commonplace.

Moreover, given the increased exposure of farmers to production income risk from market price volatility (following reduced price support), farmers themselves were becoming increasingly attracted to the certain income stream associated with participating in environmental stewardship schemes. Schemes such as the English Higher Level Stewardship Scheme became increasingly important to farmers in determining agricultural land use. It is now a common sight to see field margins and buffer strips side-by-side with crops as joint features of agricultural land use.

Figure 6.2: Hedgerows (in this picture with oak tree) can be now be counted as part of Ecological Focus Areas on farms. Farmers in the EU need to set aside a portion of their land to such uses in order to be eligible for CAP payments.

Source: Photo by Tom Hynes, CCBY-SA 3.0.

Lessons learnt

Unforeseen consequences

Set-aside was introduced as a production control instrument. However, what became clear during the 1990s was that, in being forced to set aside agricultural land, farmers were delivering enhanced environmental benefits from that land both in terms of reduced negative consequences, such as nitrate pollution and soil erosion from cropping, and in terms of increased positive consequences, such as improved habitats (see Rygnestad and Fraser 1999 for research findings in support of these consequences). That is, the reality of the compliance set-aside policy was that farmers, in taking the land out of production that was least detrimental to their production income, were also de-extensifying their land use in ways that were delivering a win–win for the levels of environmental goods and services provided by agricultural land.

This evidence set the European Commission on the path of developing an agricultural land use policy within the Agenda 2000 CAP. Although it was based on the concept of set-aside as a production-control policy, it transformed itself into a policy that saw farmers as environmental stewards, charged with the task of managing their agricultural land to provide environmental goods and services for society, and being appropriately remunerated for this provision within the CAP's budget.

The recently revealed CAP Reform 2014–2020 has seen the re-introduction of set-aside as a requirement, although it is now to be called an 'Ecological Focus Area' (initially 5 per cent of land, rising to 7 per cent in 2017).

Carrots versus sticks

The EU was quick to move its agricultural land use policy into the realms of compulsory participation by farmers (and is set to move further with the Ecological Focus Area of the CAP Reform 2014–2020). Why have other developed countries such as the US and Australia not taken this step?

I don't think this difference is to do with the history of government support for farm incomes — the US is not unlike the EU in having a long tradition of supporting farmers' incomes with taxation receipts, whereas in Australia there is only a history of taxpayer-funding for drought relief.

Rather, I think this difference is more likely to be due to different perceptions between these countries in the level of demand by society for the provision of environmental goods and services from agricultural land. These demand differences probably have their origins in the proportion of the population living in or close to the agricultural landscape, and therefore more exposed to the environmental problems created by the farming of this landscape.

As a consequence, I think the EU has felt (increasingly) socially empowered to use policy sticks on its farmers to deliver improved environmental goods and services from the agricultural landscape, whereas in the US and Australia policymakers have felt obliged to mainly offer carrots to farmers to do so.

Environmental benefits versus foregone agricultural income

The success of environmental stewardship in the UK in terms of improving the provision of environmental goods and services from the agricultural landscape has been largely based on the voluntary participation of farmers in what is called the 'Higher Level Stewardship Scheme'. In this context, two of the most popular (with farmers) scheme options are field margins, whereby a farmer leaves a 4–5 metre margin around each cropped field, and buffer strips, whereby a farmer leaves a substantial uncropped strip of land which is adjacent to a waterway, wetland, or woodland area. Farmers receive the estimated foregone cropping income from these set-aside areas, even though this is, in most cases, an overestimation of the productivity of such land.

Recent empirical research in the UK has revealed that society places higher values on some components of the overall agricultural landscape than others (particularly upland areas compared with lowland areas), and that the extent to which people benefit from agricultural landscape depends on its location — with areas of landscape closer to large population centres having higher overall social value (see, for example, Garrod et al. 2014).

As a consequence, the UK is in the process of introducing a new Countryside Stewardship Scheme, whereby applications to participate will be scored according to their fit with these empirically validated environmental benefit landscape attributes so that, although payments to farmers will still be based on foregone agricultural income, at least the cost-effectiveness of environmental stewardship in delivering benefits to society will be improved.

Acknowledgements

This chapter is adapted from chapters 1 and 17 of the author's book, *Applications of principal-agent theory to agricultural land use policy: Lessons from the European Union*, published by Imperial College Press in 2015.

References

Ervin, D.E. (1988) 'Cropland diversion (set-aside) in the US and UK', *Journal of Agricultural Economics* 39(2): 183–96.

Garrod, G., E. Ruto, K. Willis and N. Powe (2014) 'Investigating preferences for the local delivery of agri-environmental benefits', *Journal of Agricultural Economics* 65(1): 177–90.

Gasson, R. and C. Potter (1988) 'Conservation through land diversion: A survey of farmers' attitudes', *Journal of Agricultural Economics* 39(2): 340–51.

Rygnestad, H. and R.W. Fraser (1999) 'An assessment of the impact of implementing the European Commission's Agenda 2000 cereal proposals for specialist wheat growers in Denmark', *Journal of Agricultural Economics* 50(2): 328–35.

7

A brief history of agri-environment policy in Australia: From community-based NRM to market-based instruments

David Salt

Key lessons

- Australia's agri-environment policy began in the 1980s. Early measures focused on community-based NRM, increasing awareness, and building social capital through Landcare.
- The perceived success of these early efforts enabled a ramping up of government investment through the Natural Heritage Trust and successive programs.
- There have been repeated failures to demonstrate measurable outcomes from this increased investment.
- This has led to a greater focus on targeted, strategic, and accountable programs.
- The capacities required to effectively deliver these programs have been inadequately addressed.

Figure 7.1: Early measures in agri-environment policy focused on community-based NRM.
Source: Photo by Greening Australia.

A very brief history

A very brief history of agri-environment policy in Australia would read something like this: It began back in the 1980s. It started small and focused on engaging the community and building capacity. It rested heavily on a volunteer effort. Initial efforts were well received and the size of the investment grew during the 1990s. However, despite increased levels of investment, agri-environment programs failed to produce enduring environmental outcomes. This failure has led more recent programs to focus on specific environmental assets through the application of a suite of market-based instruments (MBIs).

It sounds pretty dry when summarised like this, but the bottom line is that agri-environment schemes in Australia have not improved biodiversity or reversed the ongoing degradation of land and water resources. Farming is the country's major land use, it always has been, and Australia's governments have invested significant amounts of money in programs in our agricultural landscapes — most commonly referred to in the literature as agri-environment programs — that aim to reverse these declines. To date, after spending well over $6.5 billion, this investment hasn't produced many enduring environmental outcomes (Hajkowicz 2009).

This chapter seeks to expand on this very brief history. It will discuss the rise of the agri-environment and the widespread concern that investments in this space are not working. It will also touch on the counter view that the greater targeting and the complementary use of market instruments to deliver this investment may be sacrificing some of the gains made in earlier rounds.

The emergence of the agri-environment

While the literature talks about agri-environment policy, the term 'agri-environment' is not one you'll hear much out on any Australian farm. It's a description that originated in Europe as part of the Common Agricultural Policy (see Chapter 6 by Rob Fraser). Anecdotal feedback suggests Australian farmers regard the notion of agri-environment as 'smelling a bit Euro' (conveying a widespread feeling in the Australian agricultural sector that European farming receives too much government support and protection).

In Australia, the term more commonly used is 'natural resource management' (NRM), though this term is unacceptably broad for our purposes as it covers a wide range of issues that fall well outside of the farming landscape (e.g. the management of national parks). So, in this chapter (and this book), we keep with the term 'agri-environment' and use it mainly as an adjective describing government investment in environmental programs on agricultural land. But the agri-environment is also a concept embodying the notion that our agricultural lands aren't just about the production of saleable commodities (e.g. food and fibre). They are also a space that provides a range of ecosystem goods and services that are valued by the broader public (e.g. biodiversity and rural amenity).

When did the agri-environment emerge? Well, of course, it's always been there in terms of our rural landscapes being valued for multiple purposes, not just production (Watson 2014). Indeed, over the past century the relative economic value of our agricultural lands has declined from 20 per cent of Australia's GDP in 1901 to less than 3 per cent in recent years (Hajkowicz 2009).

However, it was during the 1980s that Australian governments, and particularly the federal government, began investing considerable amounts into these landscapes to protect these multiple values, moving from a purely commodity-based industry to one that included social, economic, and environmental outcomes (Clayton et al. 2011).

Why then? There are many drivers behind this shift away from a single focus, but the 1980s was a time of many international and national conversations on conservation (with a World Conservation Strategy released in 1980, followed by a National Conservation Strategy for Australia in 1983) and sustainability (with the release of the groundbreaking Brundtland Report in 1987).

The notion of sustainability was brought to the fore particularly in Australia with a historic drought ravaging our production landscapes. The connection between land degradation and sustainable production was given a high profile. Of course, dealing with environmental degradation was not a new thing in 'the land of droughts and flooding rains', and state governments had been establishing soil conservation committees to address soil erosion from the 1930s (just as the US established the Soil Conservation Service following the dust-bowl years). But it wasn't until 1983 that the Commonwealth Government established an overarching process: the National Soil Conservation Program. This program undertook research, and provided advice and extension, with the aim of achieving cooperation between community, farmers, and government (Curtis et al. 2014). Starting from a small base, funding grew over the program's life with over $10 million being spent in its final year of 1988/99 (Hajkowicz 2009).

In addition to a growing national focus on how to deal with land degradation and conservation, the 1980s also saw community networks in the form of Landcare groups emerging in Victoria and Western Australia (with state government support). These networks facilitated education, raised awareness, and catalysed activities on the ground. Towards the end of the 1980s, an unprecedented alliance between the conservation lobby (the Australian Conservation Foundation) and the farmer lobby (the National Farmers Federation) proposed a Decade of Landcare, in which community action would raise awareness and catalyse investment to repair and nurture the land (Curtis and De Lacy 1998). The Australian Government accepted the proposal, and in 1989 announced the funding of $360 million for a Decade of Landcare. The Landcare movement emphasised local responses to landscape-wide challenges of national concern. It was hailed internationally as a remarkable initiative.

The National Landcare Program initially ran as part of the National Soil Conservation Program. Between 1990 and 1996 it was run as an independent program, and then through to 2008 as a subprogram of the Natural Heritage Trust. The National Landcare Program officially concluded in 2008, and has recently been resurrected by the current federal government. All of which reflects the patchwork history of this national program.

There was significant enthusiasm around the vision of Landcare. Local groups were encouraged to self-organise around land and water issues, with small amounts of short-term funding. The aim was to harness community spirit and catalyse greater investment from a number of sources rather than for government to provide most of the funding itself (Tennent and Lockie 2013). The approach proved popular: within five years of being established the movement had expanded from 200 local Landcare groups to 2,200 (Martin and Woodhill 1995). By 2004, there were some 4,500 groups, consisting of around 120,000 volunteers, including 30 per cent of farmers. Some surveys even recorded that farmers who did not join Landcare groups believed that their properties had benefited from participation in Landcare activities (Curtis and De Lacy 1996, and see 'Evolution of Landcare in Australia': www.agriculture.gov.au/ag-farm-food/natural-resources/landcare/publications/evolution-of-landcare-in-australia).

Ramping up investment

The National Landcare Program signalled a significant increase in investment and interest in NRM across Australia. This ramp up in investment paralleled large increases in investment in NRM in the United States and the European Union (EU), albeit Australia was spending a fraction of the amounts they were investing (Hajkowicz 2009). Around the world there was a growing societal concern about environmental degradation, much of which was (and still is) connected to the intensification of agricultural production, and societal permission for farmers to be paid for the public good of environmental outputs. In the EU and the United States, these environmental payments were offered as an alternative way to continue government support to farmers, but in a manner that didn't distort production or harm international trade.

But the situation was different in Australia. At the end of the 1980s, the Australian Government wasn't supporting farmers to the same degree as in the United States or the EU, and there was little trade protection on offer. This reflected Australia's smaller population (and therefore smaller tax base) and considerably smaller agricultural sector. Yet the actual size of our agricultural estate is comparable to these other regions.

The Australian Government simply didn't have the available resources to protect and subsidise agriculture in the manner in which agriculture was supported in the EU and the United States. In Australia's case, it wasn't a matter of attempting to wean farmers off existing support. Because there was little existing support in Australia, there was no pool of funds available, as was the case in the United States and the EU. However, there was a common growing societal concern (from both the city and the country) about the environmental costs of agriculture.

Landcare was an impressive achievement, and yet it was soon realised that building networks, raising awareness, and changing attitudes was not enough by itself. Significantly greater investment was required if Australia was to effectively tackle the multiple problems of soil erosion, water degradation, and biodiversity decline (Curtis 2000; Hajkowicz 2009).

What Landcare did demonstrate, however, was that government funding for environmental activity in regional Australia enjoyed widespread electoral support. Indeed, when the Australian Government controversially sold the national telecommunications company Telstra in 1997, it dedicated $1.3 billion from the sale towards a new national NRM program called the Natural Heritage Trust (NHT). This represented a quantum leap in NRM, received widespread community support (Curtis et al. 2014), and went some way to assuaging the electorate's concerns about the sale of a large public asset such as Telstra.

On paper it was a win–win — a means of providing support to an important constituency (the farming sector) in a politically acceptable fashion (payments for the environment) that would produce dividends for everyone (a healthier, productive landscape).

The only problem was that the promised environmental outcomes could not be measured. This was pointed out to the government by Australia's national auditors on several occasions (see ANAO 2008 for a summary). The NHT was subsequently described as suffering from the Vegemite approach (a uniquely Australian description).[1] In other words, the investment was spread too wide and too thinly to produce changes that could be readily observed.

NRM evolution

The billion-dollar NHT didn't counter the growing environmental threats facing Australia's agricultural landscapes, indeed, what it had achieved couldn't even be measured. What it did do, however, was set a benchmark for what the nation expected to be spent on environmental programs across our production zones.

Following the NHT, four national NRM programs were rolled out over the following decade: a second round of the Natural Heritage Trust (NHT2), the National Action Plan for Salinity and Water Quality (NAP), Caring for our Country (CfoC), and, most recently, a new version of the National Landcare Program. At the time of writing, this new National Landcare Program is still being formulated, so it won't be discussed here. Each program was sold as a billion-dollar (plus) fix that would halt environmental decline and ensure the productive future of our agricultural landscapes. Figure 7.2 outlines the amounts committed to each program.

---

1   For non-Australian readers, it should be pointed out that Vegemite is an Australian food product: a black, oily, salty spread for toast and bread.

The evolution of Australian NRM programs

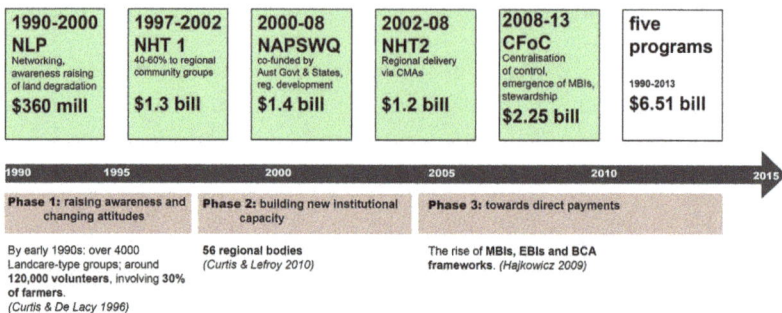

1990-2000 NLP Networking, awareness raising of land degradation $360 mill	1997-2002 NHT 1 40-60% to regional community groups $1.3 bill	2000-08 NAPSWQ co-funded by Aust Govt & States, reg. development $1.4 bill	2002-08 NHT2 Regional delivery via CMAs $1.2 bill	2008-13 CFoC Centralisation of control, emergence of MBIs, stewardship $2.25 bill	five programs 1990-2013 $6.51 bill

1990	1995	2000	2005	2010	2015

Phase 1: raising awareness and changing attitudes	Phase 2: building new institutional capacity	Phase 3: towards direct payments
By early 1990s: over 4000 Landcare-type groups; around 120,000 volunteers, involving 30% of farmers. (Curtis & De Lacy 1996)	56 regional bodies (Curtis & Lefroy 2010)	The rise of MBIs, EBIs and BCA frameworks. (Hajkowicz 2009)

Figure 7.2: The evolving focus of Australian NRM programs.
Source: Modified from Hajkowicz 2009.

Figure 7.2 summarises the five major NRM programs that have been rolled out in Australia over the past quarter of a century. Together they have resulted in a total of over $6.5 billion in expenditure. Stefan Hajkowicz frames this roll out as an evolution that proceeded through three phases (Hajkowicz 2009). He labelled these as awareness raising (phase one), building institutional capacity (phase two), and towards direct payments (phase three). This categorisation has proved quite popular in the academic literature, with several authors using it as a frame for their own examination of this period. (For excellent accounts of the history of NRM investment policy in Australia see Morrison et al. 2010; Clayton et al. 2011; and Tennent and Lockie 2013, all of which use the categorisation described by Hajkowicz).

In broad brush strokes, phase one marks the birth of community-based NRM through the establishment and growth of Landcare, and the ramp up of environmental spending to billion-dollar programs.

Phase one elevated community-based NRM to centre stage and met with strong electoral support. But it didn't address the environmental degradation that prompted it. Phase two saw an effort to respond to criticisms that these earlier investments, while changing attitudes and building social networks, weren't achieving improvements in resource conditions. Problems of salinity, soil erosion, water pollution, and biodiversity decline seemed to continue unabated. In response, a greater focus was placed on strategy and regional delivery with program funding being delivered via regional catchment management authorities (CMAs).

Phase two saw the development of regional bodies to better target the investment to regional needs while assisting in the delivery of resources in a more strategic manner.

In 2000, the Australian Government released a new NAP and in 2002 the next version of the NHT. Both were to be delivered by these regional bodies, with funding being dependent on proposals aligning with regional plans. The regional scale was considered appropriate to support NRM, as it is holistic and comprehensive (Curtis et al. 2014). It was expected that this regional empowerment would be complementary to Landcare, which still operated at the local level.

Phase three saw NHT2 being replaced by CfoC, with the Australian Government adopting a more centralised, strategic, and competitive approach to NRM. A key component of this approach involved direct payments to farmers through MBIs. For several years, Australian Governments had been experimenting with MBIs, with a focus on reverse auctions in which farmers were asked to tender for the provision of specified environmental actions and the government selected those bids that gave them greatest value for their money. (This was the manner in which money was given out for the Environmental Stewardship Program, described in Chapter 3.) Some claim MBIs are an effective and efficient way to allocate resources (e.g. Stoneham et al. 2003). Others are worried that the claims of efficacy and cost-effectiveness have not been substantiated, and that we don't yet have functioning markets for environmental services (Curtis and Lefroy 2010).

Evolution or devolution?

Viewed as a phased progression, as set out by Hajkowicz (2009), it might be said that NRM policy in Australia has grown more sophisticated and efficient over its 25-year history, harnessing a wider range of tools and players; that each phase has developed new strategies and capacities that have underpinned subsequent investments. Community awareness and social capital developed in the first phase complemented (and underpinned) growing regional capacity and strategy development in the second phase. The roll out of MBIs guided by various environmental benefit indexes in phase three builds on top of the community capacity building of phase one and regional capacity building of phase two.

Indeed, Hajkowicz's progression finishes with the promise of the delivery of a new Environmental Stewardship Program as part of CfoC that will target 'matters of significant interest' (e.g. threatened ecosystems) through direct payments to farmers which will result in contracts that last up to 15 years — significantly longer than earlier schemes and more in accord with time frames of the ecological processes the scheme is attempting to protect (see Chapter 3 by Emma Burns and colleagues for a summary of the Environmental Stewardship Program). Promoting the scheme and collecting information on the suitability of bids would be done in conjunction with the regional CMAs, and there were significant resources put aside for monitoring and evaluation of those bids that ended up receiving contracts. In many ways, this confirms the suggestion of an evolution of policy that is learning and improving with each successive round.

Yet there is a growing counter view as well, suggesting that the earlier achievements made through Landcare and the development of regional delivery have been forgotten, and that the stocks of human and social capital that were developed during these earlier times have been degraded and lost. Some have suggested that our drive for efficiency may have come at the cost of effectiveness (Curtis and Lefroy 2010). Several audit reports on national NRM programs have pointed out that a lack of monitoring of outcomes and a focus on outputs rather than outcomes has made it impossible to determine whether this new approach is working (ANAO 2008, 2014).

Evidence has emerged of declining membership in community Landcare groups in parts of Western Australia, Victoria and New South Wales. Volunteer burn out has been reported in many places, with many community Landcare groups in sleeper mode or ceasing to exist (Tennent and Lockie 2013). One of the designers of Landcare recently described the program as little more than 'a threadbare patchwork quilt of tired volunteers, waiting for the next government program with a new website, a new logo, a new departmental name, and less money than it had before' (Bush Telegraph 2014).

When NHT2 and NAP were rolled out, support for Landcare was supposed to flow through the regional CMAs: Landcare provided the grass-roots knowledge, volunteers and connections; CMAs provided the regional targets, strategy and expert knowledge. But the experience in many places was of competition and displacement rather than complementarity and cooperation.

With CfoC, the regional CMAs found their role in the delivery of community-based NRM constrained, with a reduction in base funding, increased competition with a wider range of organisations, increased transaction costs, and the need to align projects to national priorities. In some cases, regional bodies redirected support away from Landcare support, cutting the available workforce in Victoria, as one example, by more than half (Curtis et al. 2014). At the same time, state governments were making large cuts to NRM expenditure across Australia, translating to a 40–50 per cent reduction in funding to regional bodies in New South Wales and Victoria (Curtis et al. 2014).

It seemed to some observers that while the government had been devolving responsibility to regions and local groups to affect landscape change, it hadn't been passing down the authority or agency, or even the necessary resources, to underpin the required effort (Wallington et al. 2008). Indeed, given the lack of measurable outcomes from previous investment in community-based NRM, it has been suggested that governments were introducing MBIs as a way of controlling farmers' behaviour.

This raises a major tension. On the one hand, a collaborative approach is still held up as an important dimension of NRM funding and management — that investment should cultivate and work through partnerships between individuals (farmers), groups (Landcare), regions (CMAs), and government (state and federal). Indeed, such an approach is consistent with the government's neoliberal agenda of shifting responsibilities from governments to communities and individuals. However, on the other hand, this evolution has seen the original intent of empowering communities changing to one where these bodies have simply become on-ground implementation agents of strategies decided elsewhere, largely within a centralised government (Lockwood and Davidson 2010).

The baby and the bathwater

This brings us back to our very brief history of agri-environment policy. Agri-environment policy began in the 1980s as an effort to arrest declining environmental conditions in our agricultural landscapes. It took off as a partnership, and sought to empower communities and regions. But it didn't fix the things it was set up to fix, so, over the years, it progressively moved the emphasis from empowerment to targeted grants focused on specific environmental assets. It switched from building adaptive capacity and raising awareness, which is extremely hard to quantify and measure, to a focus on investing in specific actions on definable things. It moved from cultivating an ethic rooted in collaboration, sharing, and volunteerism to a culture of benefit–cost analysis and fee for service. We are much better placed to define and measure what we should be doing, but have lost sight of the suite of community capacities that enable those actions to be effectively undertaken.

Australia is currently seeing the roll-out of a new national billion-dollar program named the National Landcare Programme, which will involve planting many trees (the target is 20 million), a Green Army (to establish on-ground works while training young and unemployed Australians), and speaks the rhetoric of empowering regional Australia. Whether it succeeds or sustains our track record of failure will likely depend on whether we have the capacity to learn from our quarter of a century of effort in the agri-environment space.

References

Australian National Audit Office (ANAO) (2008) *Regional Delivery Model for the Natural Heritage Trust and the National Action Plan for Salinity and Water Quality*, Australian National Audit Office Audit Report No. 21, 2007–08. Available at: www.anao.gov.au/uploads/documents/2007-08_audit_report_21.pdf.

ANAO (2014) *Administration of the Biodiversity Fund Program*, Australian National Audit Office Audit Report No. 10, 2014–15. Available at: www.anao.gov.au/~/media/Files/Audit%20Reports/2014%202015/Report%2010/AuditReport_2014-2015_10.pdf.

Barr, N. (2009) *The house of the hill: The transformation of Australia's farming communities*, Land and Water Australia, Canberra.

Bush Telegraph (2014) 'Spirit of Landcare is "lost"', ABC Radio National. Available at: www.abc.net.au/radionational/programs/bushtelegraph/landcare-funding/5455356.

Clayton, H., S. Dovers and P. Harris (2011) *HC Coombs Policy Forum NRM initiative*, The Australian National University, Canberra. Available at: crawford.anu.edu.au/public_policy_community/research/nrm/NRM_Ref_Group_Combined.pdf.

Curtis, A.L. (2000) 'Landcare: Approaching the limits of voluntary action', *Australian Journal of Environmental Management* 7: 19–27.

Curtis, A. and T. De Lacy (1996) 'Landcare in Australia: Does it make a difference?', *Journal of Environmental Management* 46: 119–37.

Curtis, A. and T. De Lacy (1998) 'Landcare, stewardship and sustainable agriculture in Australia', *Environmental Values,* 7: 59–78.

Curtis, A.L. and E.C. Lefroy (2010) 'Beyond threat- and asset-based approaches to natural resource management in Australia', *Australian Journal of Environmental Management* 17: 134–41.

Curtis A., H. Ross, G.R. Marshall, C. Baldwin, et al. (2014) 'The great experiment with devolved NRM governance: Lessons from community engagement in Australia and New Zealand since the 1980s', *Australasian Journal of Environmental Management* 21(2): 175–99.

Hajkowicz, S. (2009) 'The evolution of Australia's natural resource management programs: Towards improved targeting and evaluation of investments', *Land Use Policy* 26: 471–8.

Lockwood, M. and J. Davidson (2010) 'Environmental governance and the hybrid regime of Australian natural resources management', *Geoforum* 41(3): 388–98.

Martin, P. and J. Woodhill (1995) 'Landcare in the balance: Government roles and policy issues in sustaining rural environments', *Australian Journal of Environmental Management* 2(3): 173–83.

Morrison, T., C. McAlpine, J. Rhodes, A. Peterson and P. Schmidt (2010) 'Back to the future?: Planning for environmental outcomes and the new Caring for our Country program', *Australian Geographer* 41(4): 521–38.

Pannell, D.J. and A.M. Roberts (2010) 'Australia's National Action Plan for Salinity and Water Quality: A retrospective assessment', *Agriculture and Resource Economics* 54: 437–56.

Stoneham, G., V. Chaudhri, A. Ha and L. Strappazzon (2003) 'Auctions for conservation contracts: An empirical examination of Victoria's BushTender trial', *Australian Journal of Agricultural and Resource Economics* 47: 477–500.

Tennent, R. and S. Lockie (2013) 'Vale Landcare: The rise and decline of community-based natural resource management in rural Australia', *Journal of Environmental Planning and Management* 56(4): 572–87.

Wallington, T., G. Lawrence and B. Loechel (2008) 'Reflections on the legitimacy of regional environmental governance: Lessons from Australia's experiment in natural resource management', *Journal of Environmental Policy and Planning* 10(1): 1–30. DOI:10.1080/15239080701652763.

Watson, D. (2014) *The Bush: Travels in the heart of Australia*, Penguin Books, Ringwood.

Part II.
The birds and the beef

8

Can recognition of ecosystem services help biodiversity conservation?

Saul Cunningham

Key lessons

- Ecosystem services thinking explicitly brings farmers and their activity into the framework for decision-making and provides a model communicating the benefits of nature conservation that is effective for some audiences.

- Communicating the benefits of ecosystem services to landholders can promote the advantages of nature conservation actions in their landscapes, increasing adoption and community support.

- We should not assume that ecosystem services and biodiversity conservation always pull in the same direction for land use decision-making.

- Ecosystem services thinking can help to identify stakeholders and beneficiaries in a way that improves policy design.

- Management to improve ecosystem services sometimes requires a deeper understanding of ecosystems than we currently have.

Nature conservation policies have increasingly invoked ecosystem services as part of their rationale. What were formerly biodiversity polices are now increasingly communicated as policies for biodiversity and ecosystem services. Views on this shift range from those who argue that ecosystem services thinking is not particularly useful to conservation policy (e.g. Srivastava and Vellend 2005) through to those who argue that the vision is beginning to have real impact (e.g. Daily et al. 2009). Agri-environment schemes are particularly pertinent in this debate, because they target conservation outcomes in production-oriented landscapes. In this chapter, I examine the extent to which incorporating ecosystem services concepts into policy can affect both the targeting of outcomes and encouraging uptake and engagement in agri-environment schemes.

Ecosystem services thinking

The essential feature of ecosystem services thinking is to explicitly focus on the values that people derive from nature, unashamedly placing humans at the centre of things. In this way it contrasts to philosophies that emphasise the inherent values of nature, in particular the idea that nature has a right to exist, or that it has an intrinsic value that humans should act to preserve. The boundary between these perspectives can be messy because some ecosystem services frameworks include cultural values. Rather than being drawn into the literature on definitions, here I will use 'ecosystem services' thinking to denote frameworks that put more emphasis on the ways in which nature serves people with things we use (e.g. food and water) or processes that benefit us (e.g. crop pollination, pest control, nutrient cycling). This is in contrast with classical biodiversity conservation thinking, which aims to maximise the preservation of species and genetic diversity.

Combining ecosystem services thinking with biodiversity conservation thinking is relatively simple if the two are tightly linked. If protecting biodiversity is the best route to protect ecosystem services, then the two goals would appear to be extremely compatible. Sure enough, the pattern from extensive reviews of the literature indicate that, in general, more diverse communities tend to support better ecosystem services (e.g. Cardinale et al. 2006), especially if one considers stability over time in the assessment (simplified systems might have high levels of provision in some circumstances, but do not respond well to perturbation).

One fundamental rationale for adopting a nature conservation strategy hinges on the idea of protecting these ecosystem services; in this way, the ecosystem services perspective is closely aligned with the idea of sustainability. However, it is not always the case that the goals of biodiversity conservation and ecosystem services protection are always and everywhere in step (Macfadyen et al. 2012). If, for example, one were to assess a landscape and design a system of protected areas to maximise conservation of biodiversity, and then compare it with a system that optimised for water yield, carbon sequestration, and pest control (three example ecosystem services), one would no doubt produce different protected area networks. This is because, at the landscape scale, a focus on species conservation pulls in different directions to a focus on services, and one ecosystem service might pull in a different direction to another.

Ecosystem services in agricultural landscapes

Conservation goals and strategies depend on the landscape in question. Agricultural landscapes are characterised by significant historical loss of biodiversity and great emphasis on utilitarian production values (i.e. the provision of food and fibre). It is in this context that ecosystem services thinking will be most likely to influence the uptake of conservation actions and the design of strategies. Indeed, pure biodiversity-focused strategies will often downgrade the relevance of conservation in agricultural landscapes, compared with other landscapes which are less modified by people, because much native biodiversity is already lost. It is important, though, to recognise that ecosystem services are relevant to farmers in three quite distinct ways (see Figure 8.1). First, farmers manage their land to provide one of the key ecosystem services to broader society (i.e. food and fibre). Protection of this service means maintaining farming in the landscape. Second, agricultural practice requires the support of a raft of ecosystem services. The benefit of these services flows primarily to the farmer, and second to society that uses the food and fibre.

Third, farmers have the potential to manage their land for a range of ecosystem services for the benefit of broader society rather than for themselves — by sequestering carbon and thereby reducing the risk of climate change, for example. If ecosystem services thinking is going to be used to shape agri-environment schemes, the different kinds of benefit flow described in this framework need to be understood by policymakers.

Figure 8.1: Wasps help to control pests in agricultural systems and pollinate crops and native species.
Source: Photo by Saul Cunningham.

Reflecting on the fact that farmers provide society with food and fibre is important in terms of validating the role of agriculture in the landscape, and recognising the value of that land use to society. But it does not follow that all agricultural practice is justified by the need for food and fibre. Some conservation planning exercises include production in a cost–benefit analysis, asking if the cost of a conservation action (in terms of lost production) is justified in terms of the benefit (in terms of nature saved) (e.g. Hodgson et al. 2010). This approach leads into the land sparing/land sharing spectrum of choices, discussed in Chapter 9. The pragmatic reality is that land use decisions will always be influenced by use values, and so the production side of the equation will always affect the decision, even if it is not in a formal cost–benefit analysis. The effectiveness of payments to landholders to effect land use change will, of course, depend on the production values of land.

Recognising and communicating benefits

Ecosystem services that support agricultural activity can be considered private benefits (see chapters 14 and 18) because there is a direct economic pay-off if farmers increase them. For example, if protection of native grasslands around a field margin increases pest control and crop pollination on farm, this would be a private benefit to the farmer, albeit linked to nature conservation. But there are great challenges in using these kinds of benefit to motivate land use change (see Chapter 12). In most cases, farmers will not have enough information to guide practice change to exploit these benefits. Often the benefit of improvement in one particular ecosystem service will not be great enough to justify the cost (e.g. lost area of production). The benefit may also require land use changes on a scale beyond the area of a single farm, and therefore beyond the scope of a single landholder's decision-making. For example, benefits of natural pest control might be influenced by non-crop vegetation that is kilometres away, and therefore require cooperation among neighbours to achieve. There may also be a lot of uncertainty around the reliability in supply of the ecosystem service over time and space compared with the relative certainty associated with agronomic inputs, invoking new risks.

For these reasons, the use of ecosystem services strategies to increase the adoption of on-farm nature conservation requires landholders and policymakers to have a high level of understanding of the system, and an honest assessment of the economic challenges facing farmers. In many circumstances, ecosystem services benefits may only partially offset the costs of nature conservation practices. The nature and magnitude of costs and benefits are likely to vary according to the economic and biophysical characteristics of the type of production system (e.g. annual cropping versus grazing). Benefits are likely to be greatest when combined with strategic assessment of the productivity of different parts of the farm, so that poor areas for production are relinquished but ecosystem services benefits are realised. Public policy has a potentially significant role in helping farmers to make the changes that lead to these private benefits from ecosystem services.

Ecosystem services that farmers, as land managers, provide to society can be considered public benefits (see Figure 8.1). Farmers are unlikely to make costly changes to farm practice simply to support benefits to broader society, unless they are encouraged by incentives (see Chapter 4). In most cases, however, land use change in the interest of public benefit (such as management of streamside vegetation to support clean water downstream) will also influence nature conservation outcomes and sometimes even private benefits, such as shade and shelter for stock. In practice, one needs to assess all these dimensions. One of the strengths of the ecosystem services paradigm is in making sense of these private and public benefits in a way that can support cost-effective investments.

The social acceptability of land management action is likely to be greatest if advocates are effective in communicating the multiple benefits — private benefits, public benefits, and nature conservation benefits — associated with the change. Participation in native vegetation management programs in Australia was motivated by a mix of drivers, including ecosystem service benefits to the farm along with nature conservation (see Chapter 12). Payments for stewardship or, for example, carbon farming will attract some landholders who are sympathetic to the program, even if the payments are modest. But one can expect wider acceptance of these programs (even at the same payment level) if there is an understanding that the land use changes will provide private ecosystem services benefits to the farmer in addition to the ecosystem service and biodiversity conservation outcomes recognised by broader society (Lin et al. 2013). The idea of recognising these co-benefits is particularly relevant for the relationship between biodiversity conservation and ecosystem services, because while they are not necessarily jointly optimised by the same actions, any given action for one is likely to lead to a marginal benefit in the other, compared to the alternative scenario of no land use change. Well targeted incentives, such as payments for fencing, might play an important role in supporting landholders to take up these opportunities.

Figure 8.2: A wheat crop amongst bushland.
Source: Photo by Belinda Gibson.

To this point, I have argued that an ecosystem services perspective has the potential to drive better engagement in conservation policies in agricultural landscapes, but also that there are substantial knowledge gaps. In particular, it is difficult to argue that there are benefits to farmers, or even to broader society, unless we have a good understanding of the natural systems we are dealing with. We have enough understanding to promote some general principles that suggest biodiversity and ecosystem service benefits are supported by protection of habitat remnants. For example, conserving patches of native vegetation in agricultural landscapes is known to support crop pollinators (Garibaldi et al. 2011). But it is more challenging to advise a landholder on the specific benefits that can be expected to flow from a given investment in ecological restoration, and by how much and under what circumstances these services will improve a farmer's bottom line. The outcome considering multiple benefits is more likely to be positive than that from a single ecosystem service, but also more complex to determine (Olschewski et al. 2010). It is also important to remember that while the ecosystem services perspective promotes engagement with some people, there are others who are put off by the utilitarian perspective of nature. Ecosystem services ideas should be used strategically as a means of communicating some values from nature, not as a replacement to pre-existing frameworks for nature conservation.

Paying for ecosystem services

A large part of the literature on ecosystem services focuses on economic valuations of ecosystem services and the conceptualising and design of markets or payment systems, so that money flows from the beneficiaries of the service to those who pay the cost of land management. While this has attracted a lot of attention, conservation strategies in Australia have not yet been substantially influenced by payments for ecosystem services programs. In this respect, talk appears to have outstripped action. But this does not mean that ecosystem services thinking has no relevance to nature conservation strategies. Payments for carbon sequestration are still on the political agenda and have the potential to affect nature conservation in some agricultural landscapes. It is wrong to judge the impact of ecosystem services thinking by focusing only on the degree to which payment strategies have been implemented. The best nature conservation strategies should address the sustainability of human well-being, as well as conservation of biodiversity per se, and this broader goal can be guided by ecosystem services frameworks. A focus on the utilitarian values of nature is required to get the best engagement from the landholders that manage most of the landscape, especially in agricultural landscapes (Goldstein et al. 2012).

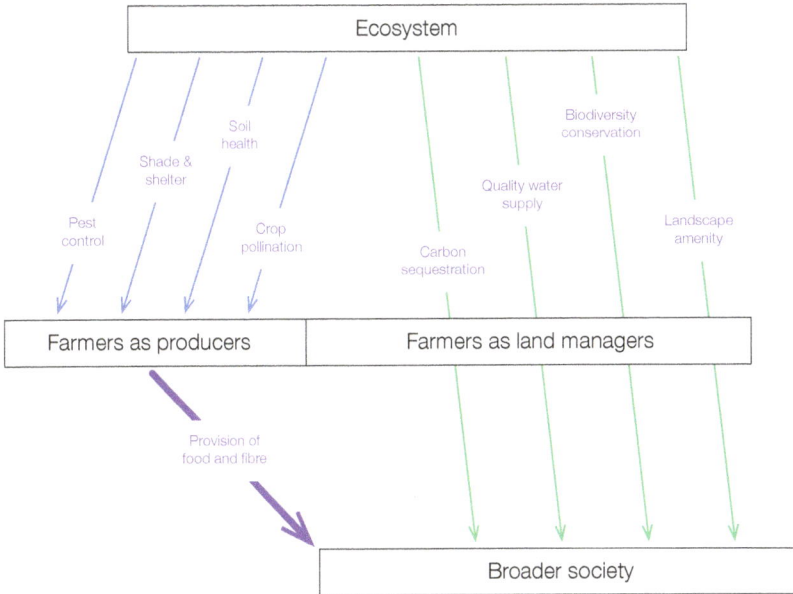

Figure 8.3: This framework highlights ecosystem services that are most relevant to farmers.

Source: Author's research.

Note: Purple text names ecosystem services that are widely recognised in the general literature. The arrows indicate how these services flow from ecosystems to human society in general, and what role farmers have in this flow. The framework deliberately separates two different roles that farmers play: their primary role as producers, and their secondary role as land managers. The large purple arrow highlights the obvious major ecosystem service: the provision of food and fibre. The blue arrows highlight the services that ecosystems provide to farmers, supporting their production activity (i.e. private benefits). The green arrows identify services that ecosystems provide to society in general (i.e. public benefits), which farmers can influence substantially in their role as land managers.

References

Cardinale, B.J., D.S. Srivastava, J.E. Duffy, et al. (2006) 'Effects of biodiversity on the functioning of trophic groups and ecosystems', *Nature* 443: 989–92.

Daily, G.C., S. Polasky, J. Goldstein, et al. (2009) 'Ecosystem services in decision making: Time to deliver', *Frontiers in Ecology and the Environment* 7: 21–8.

Garibaldi, L.A., I. Steffan-Dewenter, C. Kremen, et al. (2011) 'Stability of pollination services decreases with isolation from natural areas: A global synthesis', *Ecology Letters* 14: 1062–72.

Goldstein, J.H., G. Caldarone, T.K. Duarte, et al. (2012) 'Integrating ecosystem-service tradeoffs into land-use decisions', *Proceedings of the National Academy of Sciences USA* 109: 7565–70.

Hodgson, J.A., W.E. Kunin, C.D. Thomas, et al. (2010) 'Comparing organic farming and land sparing: Optimizing yield and butterfly populations at a landscape scale', *Ecology Letters* 13: 1358–67.

Lin, B.B., S. Macfadyen, A.R. Renwick, et al. (2013) 'Maximizing the environmental benefits of carbon farming through ecosystem service delivery', *BioScience* 63: 793–803.

Macfadyen, S., S.A. Cunningham, A.C. Costamagna and N.A. Schellhorn (2012) 'Managing ecosystem services and biodiversity in agricultural landscapes: Are the solutions the same?' *Journal of Applied Ecology* 49: 690–4.

Olschewski, R., A-M. Klein and T. Tscharntke (2010) 'Economic trade-offs between carbon sequestration, timber production, and crop pollination in tropical forested landscapes', *Ecological Complexity* 7: 314–19.

Srivastava, D.S. and M. Vellend (2005) 'Biodiversity-ecosystem function research: Is it relevant to conservation?', *Annual Review of Ecology Evolution and Systematics* 36: 267–94.

9

A perspective on land sparing versus land sharing

Anna Renwick and Nancy Schellhorn

Key lessons

- Agri-environment schemes have the potential to increase biodiversity in productive landscapes.

- The land sparing/land sharing framework uses trade-offs between agricultural yield and biodiversity to choose between two ways of achieving biodiversity conservation in agricultural landscapes, but its simplicity has generated considerable debate.

- Land sharing and land sparing represent a false dichotomy. Strategies for biodiversity conservation in agro-ecosystems form a continuum between these two extremes.

- Using different measures of diversity, considering the appropriate scale, and incorporating land use history and social factors will enable more robust management decisions to be made which best support biodiversity and production in agro-ecosystems.

- Addressing these gaps in the current production/biodiversity trade-off will enable more efficient management plans to be implemented which are directly applicable for the design of agri-environment schemes.

Introduction

Loss of biodiversity in agricultural landscapes is an important issue in conservation biology. Simultaneously, there is much concern over the ability to produce sufficient food to feed the growing global human population. There is often a disparity in the agendas of agriculturists and conservationists, with the former focusing on increasing production, often to the detriment of the environment, and the latter focusing on biodiversity conservation, with little interest in increasing food production (Foley et al. 2011). Agri-environment schemes, where farmers are paid through agri-environment subsidies to manage land primarily for wildlife, have been implemented in an attempt to conserve and prevent further declines in biodiversity on farmland. However, the effectiveness of these schemes, in terms of biodiversity conservation, has been questioned (Kleijn et al. 2011). They may also lead to reductions in crop yield (Kaphengst et al. 2010).

Scientists are currently trying to address the disparity between production and biodiversity by analysing trade-offs between agricultural yield and biodiversity conservation (Green et al. 2005; Phalan et al. 2011; Hulme et al. 2013), called Land Sharing Land Sparing (LSLS). Under this approach, land use is categorised as:

- Land sparing: the intensification of production to maximise agricultural yield within a fixed area and dedicating other land to biodiversity conservation; or

- Land sharing (also called 'wildlife-friendly farming', such as that seen within the agri-environment schemes): the aim here is to maintain biodiversity within less intensively farmed agricultural landscapes.

This conceptual framework has stimulated considerable debate, with many arguments supporting both LSLS strategies (Fischer et al. 2008; Tscharntke et al. 2012; Fischer et al. 2014; von Wehrden et al. 2014). In this chapter, we provide a perspective on the main issues of LSLS that cause debate and suggest potential improvements in evaluating the best management strategy for successfully protecting biodiversity without compromising production. Our aim is to help guide the design of future agri-environment schemes.

Figure 9.1: An example of land sharing in southern Australia, illustrating the coexistence of agriculture and biodiversity.
Source: Photo by Dean Ansell.

Issues causing debate in the LSLS framework

LSLS is a false dichotomy

The LSLS framework categorises agricultural land as one of two extremes. However, in reality there is a need to spare land from agriculture (e.g. areas of high biodiversity value and endemism), protect high value agricultural land, and identify interventions on agricultural land that will support biodiversity and contribute to ecosystem services. Agri-environment schemes typically fit into the last category. Biodiversity, such as native insect pests, pollinators, and below ground invertebrates, underpins a wide variety of ecological goods and services which contribute to agricultural productivity. Some production systems and landscapes may have the opportunity to capture many of the ecosystem services provided by biodiversity, such as pollination (Aizen et al. 2009) and pest control (Crowder et al. 2010), while other systems are more constrained to artificial inputs, such as fertiliser and lime. The issue is not whether we should spare or share land, but rather the need to identify opportunities to increase agricultural production while minimising the negative impacts on the environment and stemming further biodiversity loss — for example, managing ecosystem services in agricultural landscapes to maintain or enhance yield level rather than relying on external inputs (Bommarco et al. 2013).

Efficient allocation of land

Recent articles using the LSLS framework attempt to frame the central point of interest as food security or food production (e.g. Phalan et al. 2011; Hulme et al. 2013), however, their focus can more appropriately be described as the most efficient allocation of land. This mismatched claim has attracted considerable criticism, as food security is a much more complex concept than the simple amount of food that is produced, and agricultural landscapes are often used for much more than food production (e.g. biofuels and fibre), which needs to be considered when managing landscapes (Fischer et al. 2014). This land allocation question is central to agri-environment schemes: how much land should be designated to production and how much to protecting biodiversity, in addition to the spatial arrangement of these activities?

Biodiversity in agro-ecosystems

Different taxa respond differently to changing land use and processes, and the response by species at a plot or field level may be very different from the response seen across landscape scales. Extrapolating across species groups and various scales commonly produces misleading results. This has been seen in the current agri-environment schemes, where targeting management to preserve one species may not provide a solution for other aspects of biodiversity (Kleijn et al. 2011). Additional measures of diversity are therefore needed to ensure actions support a multitude of species rather than just a single species. Von Wehrden et al. (2014) propose measuring beta and gamma diversities instead of simply alpha diversity to better capture the landscape scale variance in biodiversity in relation to land use changes.

Figure 9.2: A bee pollinating a lupin flower on a farm in western NSW.

Source: Photo by Carl Davies (CC BY 3.0).

Scale, landscape composition and leakage

Opportunities identified to increase production from intensification need to explicitly consider scale. This will help identify the appropriate study and management scale. Many biodiversity-yield interactions are affected by the scale and composition of the landscape, yet the LSLS framework does not incorporate scale, which can lead to confusion when trying to determine appropriate management. In addition, land use is often driven by distant drivers, such as global markets, and actions taken at the farm or regional scale can impact areas further afield. This leakage affect is rarely, if ever, considered, yet can have large impacts globally, making it an additional aspect to factor in when making land use decisions, including the operation of national agri-environment schemes (Renwick et al. 2015).

Land use history

Knowing the historical context of land use can influence the current pattern of biodiversity and appropriate management. Most studies present a snapshot in time, failing to account for the history of the landscape, despite land use history having a key influence on current species diversity and land use (von Wehrden et al. 2014). The key to determining the optimum land use strategy for a landscape depends largely on if it is a frontier landscape, where primary habitat is under pressure from agriculture — for example, in northern Australia and areas of Brazil — or a traditional landscape, where farming has been carried out for centuries. Biodiversity in traditional farming landscapes tends to be concentrated in specific areas such as remnant native vegetation and plantings (e.g. shelter belts). Protection of these areas is essential. However, they are also prime target areas for conversion under intensification. In contrast, in frontier landscapes, biodiversity is more evenly distributed across the landscape and the effect of intensification is more uniform across the landscape. Land sparing may be possible in frontier landscapes, whereas land sharing is most likely in the former (von Wehrden et al. 2014).

Figure 9.3: Remnant vegetation surrounded by wheat crops in New South Wales, demonstrating a land sparing approach. Agriculture is intensified into areas separate from those specific for nature.
Source: Photo by Gregory Heath (CC BY 3.0).

Social values

Social values are unaccounted for in the LSLS framework. The disconnect between people and nature is often cited as a prime reason for a decline in biodiversity (Fischer et al. 2014). Both land sparing and land sharing may encourage the connection between nature and people, but in different ways (wild nature in protected areas, or tamer nature within farms themselves). Despite the relevance of this observation, to date no data on these social elements have so far been incorporated to the LSLS framework. Incorporating social values is essential to achieving successful conservation actions, whether land sparing or sharing is identified as being the best management option. Some actions within agri-environment schemes have social as well as conservation value — for example, hedges. Identifying measures to incorporate this in evaluating their overall value is essential.

Conclusion

Identifying the most appropriate land use management is clearly imperative for protecting biodiversity in agricultural landscapes. The LSLS framework has identified some of the major challenges and highly contested aspects of land use. However, by limiting land use decisions to two options at either of end of the intensity spectrum, the framework omits to consider land use between the extremes. Agri-environment schemes may be considered a land sparing approach, but determining how these should be designed and implemented is imperative. Here we have identified some gaps in the LSLS framework that still need to be addressed, and suggested the data that is needed to fill these gaps to enable solutions to be found. Addressing the above points will enable a more robust analysis and the development of a successful land management strategy (e.g. agri-environment scheme) to be implemented in agricultural landscapes where biodiversity can be better protected with minimal if any compromise on productivity.

References

Aizen, M.A. and L.D. Harder (2009) 'The global stock of domesticated honey bees is growing slower than agricultural demand for pollination', *Current Biology* 19(11): 915–18.

Bommarco, R., D. Kleijn and S.G. Potts (2013) 'Ecological intensification: Harnessing ecosystem services for food security', *Trends in Ecology and Evolution* 28(4): 230–8.

Crowder, D.W., T.D. Northfield, M.R. Strand and W.E. Snyder (2010) 'Organic agriculture promotes evenness and natural pest control', *Nature* 466: 109–12.

Fischer, J., D.J. Abson, V. Butsic, et al. (2014) 'Land sparing versus land sharing: moving forward', *Conservation Letters* 7(3): 149–57.

Fischer, J., B. Brosi, G.C. Daily, et al. (2008) 'Should agricultural policies encourage land sparing or wildlife-friendly farming?', *Frontiers in Ecology and the Environment* 6: 382–7.

Foley, J.A., N. Ramankutty, K.A. Brauman, et al. (2011) 'Solutions for a cultivated planet', *Nature* 478: 337–42.

Green, R.E., S.J. Cornell, J.P.W. Scharlemann and A. Balmford (2005) 'Farming and the fate of wild nature', *Science* 307: 550–5.

Hulme, M.F., J.A. Vickery, R.E. Green, et al. (2013) 'Conserving the birds of Uganda's banana–coffee arc: Land sparing and land sharing compared', *PLOS ONE* 8: e54597.

Kaphengst, T., S. Bassi, M. Davis, et al. (2010) *Taking into account opportunity costs when assessing costs of biodiversity and ecosystem action*, Ecologic Institute, Berlin.

Kleijn, D., M. Rundlof, J. Scheper, et al. (2011) 'Does conservation on farmland contribute to halting the biodiversity decline?', *Trends in Ecology and Evolution* 26: 474–81.

Phalan, B., M. Onial, A. Balmford and R.E. Green (2011) 'Reconciling food production and biodiversity conservation: Land sharing and land sparing compared', *Science* 333: 1289–91.

Renwick, A.R., O. Venter and M. Bode (2015) 'Reserves in context: planning for leakage in protected areas'. *PLoS ONE* 10(6): e0129441.

Tscharntke, T., Y. Clough, T.C. Wanger, et al. (2012) 'Global food security, biodiversity conservation and the future of agricultural intensification', *Biological Conservation* 151: 53–9.

von Wehrden, H., D.J. Abson, M. Beckmann, et al. (2014) 'Realigning the land-sharing/land-sparing debate to match conservation needs: Considering diversity scales and land-use history', *Landscape Ecology*, 29(6): 1–8.

10

Restoring ecosystem services on private farmlands: Lessons from economics

Md Sayed Iftekhar, Maksym Polyakov
and Fiona Gibson

Key lessons

- Biological conservation attempts to preserve and maintain existing habitat, while ecological restoration attempts to reverse an environmental degradation process.

- The higher cost per unit area (or per ecological outcome) to implement restoration projects, compared with conservation projects, could negatively influence their formation and acceptance.

- Broad support for restoration projects can be difficult to achieve, due to people's loss aversion behaviour.

- Uncertainty in expected biodiversity benefits can influence the acceptance and success of restoration projects.

- Social value could influence the objectives of restoration projects; the more aligned the social and environmental objectives are, the higher the chances of acceptance.

- Some of the biophysical and social benefits of restoration projects could be privately captured, which could increase acceptance of restoration projects.

- Economic incentives, such as monetary benefits, can play a crucial role in motivating private landholders to participate in agri-environment schemes, but may not be sufficient.

Clearance of the natural environment for farming and intensification of land use on existing farmlands puts pressure on the remaining natural environment (Michael et al. 2014). As a result, there is a decline of species richness as well as biotic homogenisation, as species with high conservation concern are gradually replaced by species with lower conservation concern (Donald and Evans 2006). Protecting and restoring biodiversity on private farmlands can therefore play an important part in provision of ecosystem services.

Governments have recognised the importance of conserving and restoring ecosystem services on private lands through agri-environment schemes. Agri-environment schemes have been developed in Australia (e.g. Bush Tender), the US (e.g. Conservation Reserve Program), and the European Union (e.g. Common Agricultural Policy). Under these programs, landholders receive government support in exchange for undertaking environmental management actions, such as ecological restoration.

Traditionally, ecological restoration activities and projects have been targeted, prioritised, and planned using ecological considerations (Aronson et al. 2006). Only recently have ecologists begun to include economic and social considerations in the design of restoration projects (Blignaut et al. 2014). With few exceptions, ecological restoration studies that include economics focus only on the cost side of restoration projects (Bullock et al. 2011; Wilson et al. 2012). Incorporating proper cost estimates in benefit–cost analysis is essential to make sound economic decisions in terms of the selection and prioritisation of restoration projects. However, economics can contribute more broadly to restoration programs. Lessons from economics could assist in the conceptualisation and planning of programs, reveal factors affecting program acceptance and uptake, and inform the design of restoration programs to increase success. In this chapter, we summarise some of the key lessons from economics that we hope will help improve the effectiveness and efficiency of ecological restoration and agri-environmental schemes.

Comparing costs from ecological restoration and conservation projects

Restoration projects, which are primarily aimed at restoring ecosystems of varying levels of degradation, are generally more costly than conservation projects, which are aimed at the protection of existing ecologically intact ecosystems (Blignaut et al. 2014). Some types of costs are common for both restoration and conservation projects (e.g. opportunity and maintenance costs). However, establishment costs are likely to be much higher for restoration projects, which require a variety of restoration actions with different levels of intensity (Hobbs and Cramer 2008). These actions range from allowing the ecosystem to recover without human interference — commonly referred to as passive restoration — to actively undertaking a restoration process. In agricultural landscapes, ecological restoration usually requires active intervention (Polasky et al. 2005).

The opportunity cost of ecological restoration is often much higher than the opportunity cost of creating a reserve in an undisturbed natural landscape, for the following reasons:

- Restoration projects tend to occur on intensive production, higher-value land, such as agricultural or mining landscapes, whereas conservation projects are usually located on lower-value land, such as grazing land.

- Restoring ecosystems on land impacted by intensive production is often more costly than on land used for other less-intensive purposes due to a greater rate of land use modifications.

- Restoration is an intensive process and restoration projects tend to be of a smaller scale than conservation projects, which increases both the cost per hectare of restoring small sites and the overhead costs associated with managing many smaller projects.

- Ecological restoration is undertaken when reservation of intact habitat is not an option. Such landscapes tend to be highly fragmented, which could also influence ongoing maintenance costs of restoration projects (Lindenmayer et al. 2002; McBride et al. 2010).

The success of ecological restoration projects is determined by their ecological proximity and connections to existing habitats. Ecological synergy benefits from the existing protected areas influence the capacity of the restored sites to generate ecosystem services (Bennett et al. 2006; Manning et al. 2006). Therefore, the spatial arrangement and characteristics of existing habitats inside and outside the planning areas need to be carefully considered. Otherwise, restoration projects are likely to be less effective and require higher operation costs to generate the same amount of environmental outcome (Thomson et al. 2009).

Support for restoration projects can be difficult to obtain due to people's loss aversion behaviour

Gaining social support for restoration projects can be difficult, compared to conservation projects, due to people's loss aversion behaviour. It has been observed that people are more sensitive to changes seen as losses than to gains of the same magnitude (Tversky and Kahneman 1992). For example, let us assume that an individual owns $100. According to this theory, she will suffer greater dissatisfaction if $10 is taken away from her than the corresponding level of satisfaction if she was given $10. In the presence of loss aversion behaviour, people may be more inclined to support a conservation project that proposes to protect a bushland (preventing it from being lost) than a restoration project that would restore a bushland of the same area and quality (which could mean losing productive land). Uncertainty in expected environmental outcomes of a restoration project would increase people's loss aversion behaviour.

Uncertainty in expected ecological outcomes can influence the adoption and success of restoration projects

When designing conservation programs, decision makers often have a target of what they plan to conserve or protect, even though in most cases the target is poorly defined. With restoration programs, the target is a desired environmental outcome (Yoshioka et al. 2014). These predictions are based on ecological and biophysical models, such as tree growth models or species distribution models. Even if these models use the best available data and advanced modelling techniques, there can be substantial uncertainty in their predictions (Haila et al. 2014). The probability that a restoration project will deliver the expected outcomes depends not only on the on-ground actions but many other factors, such as environmental condition, restoration technique, and socio-economic factors (Lindenmayer et al. 2002; Maron et al. 2012; Raymond and Brown 2011). Failure to take account of these uncertainties diminishes the benefits of ecological restoration. However, so far, the probability of success of restoration projects has rarely been factored into the ecological-restoration planning process (Dorrough et al. 2008).

Social value could influence the objectives of restoration projects

The objectives of restoration projects are likely to be influenced by social values. Values for conservation projects are easier (more certain) to obtain because what is there, in terms of biodiversity, is known, and the probability of conservation project failure is arguably lower than a restoration project (Robbins and Daniels 2012). For restoration, the final condition is not known due to many uncertainties. For this reason, it is often difficult to set scientific goals for restoration projects. Therefore, restoration efforts require an understanding of community values and preferences, which are highly context-dependent (Shindler et al. 2011) and should necessitate a participatory process to identify goals and aspirations for the site (Schaich 2009; Schultz et al. 2012). For example, Alam (2011) found that people's willingness to pay for restoration of a river in Bangladesh varies with their proximity to resources, their length of residence in the area, and their depth of experience with the area. In Japan, Mitani et al. (2008) found that individuals with strong environmental attitudes, a history of past visitation, and high income are more likely to prefer restoration projects to the status quo. They also observed that people with a better understanding of ecological features were willing to pay more to avoid the extinction of species. Rogers (2013) has shown that public and expert opinion about restoration efforts can diverge, especially when the public have a limited understanding and knowledge. When setting the objectives of restoration programs, there is a risk of misalignment of social and environmental objectives if governments rely only on scientific values.

Some benefits from restoration projects could be privately captured, which would then facilitate acceptance of restoration projects

Restored ecosystems can generate ecosystem services, both on-farm and off-farm. Off-farm ecosystem services include changes in water and air quality, and protection of endangered species. On-farm ecosystem services include reduction in soil erosion, shelter for grazing animals, improved soil fertility, and increased recreational benefits. The benefits of these services are captured by the participating landholders (Polyakov et al. 2015). There is evidence to suggest that some landholders may receive personal non-use values from undertaking restoration activities. For example, Greiner and Gregg (2011) found that, in some cases, farmers are more strongly motivated by stewardship aspirations than economic and social goals. Three motivation factors — economic/financial, conservation and lifestyle, and social — were found to motivate grazier management in Northern Australia. Jellinek et al. (2013) found large-scale revegetation on agricultural properties was more likely to occur where off-farm income is available and/or where the landholder has a preference to achieve environmental rather than production goals.

Understanding the preferences and behavioural dynamics of landholders is important to the success of ecological restoration activities on private land. Landholders who receive high private benefits (including non-monetary benefits) from ecological restoration would be more willing to participate in programs with lower rates of government support or incentives, which could ultimately reduce their public cost and increase their cost-effectiveness (see Chapter 14).

Economic incentives can play a crucial role in motivating private landholders to participate, but may not be sufficient on their own

There are many studies providing evidence on the importance of economic incentives in facilitating the adoption of conservation practices by landholders (see Chapter 21). However, economic incentives are not the only drivers of adoption. As Greiner and Gregg (2011) observe, it is not enough to simply pay farmers to undertake restoration activities. Factors such as resource constraints, a lack of external support, uncertainty over the future of the property, and a lack of industry cooperation can inhibit adoption of conservation practices. The influence of these factors is likely to vary between different types of landholders (e.g. lifestyle or commercial farmers) and between individuals (Pannell and Wilkinson 2009; Welsch et al. 2014). For example, Welsch et al. (2014) found that lifestyle landholders have more property in woody vegetation than dairy and beef/sheep landholders, and were more likely to increase the amount of vegetation in the future. Although lifestyle landholders may be more willing to participate in conservation practices for non-economic reasons, Pannell and Wilkinson (2009) caution that the learning and transaction costs are likely to be higher for this type of landholder.

Pannell (2008) provides a useful framework to help policymakers select the most effective policy mechanism to encourage change by landholders (see Chapter 18). This framework includes economic mechanisms (e.g. paying for services and regulation of activities) and non-economic mechanisms (e.g. new technology and extension). As noted by Pannell (2008), extension plays an important role as a policy mechanism, and should be applied when public and private net benefits are high.

Conclusion

The issue of restoring ecosystem services on private lands presents a unique setting due to the high cost of implementation, high levels of uncertainty in environmental outcomes, and high probability that the program objectives are influenced by social values and the impact of people's loss aversion behaviour. Decision makers need to be aware of these issues and take them into account when planning and implementing restoration projects. For example, by identifying landholder's preferences for different restoration objectives, it might be possible to match landholders with restoration programs that fit their preferences, facilitating the acceptability of the restoration program. Being able to direct restoration programs towards a group of landholders with high environmental motivations could help conservation agencies to deliver restoration projects more cost-effectively, as they might be willing to deliver environmental services at lower costs — although it is worth noting the potential issues with the equity of how funds are dispersed. Careful applications of these and similar lessons will help improve the efficiency and uptake of restoration projects in agri-environmental schemes.

Acknowledgements

The authors would like to acknowledge funding and support received from the National Environmental Research Program and the Centre of Excellence for Environmental Decisions. We thank Caroline Mitchell for proofreading the document. We would also like thank the reviewers and editors of the book for their constructive comments on the chapter.

References

Alam, K. (2011) 'Public attitudes toward restoration of impaired river ecosystems: Does residents' attachment to place matter?', *Urban Ecosystems* 14: 635–53.

Aronson, J., A.F. Clewell, J.N. Blignaut and S.J. Milton (2006) 'Ecological restoration: A new frontier for nature conservation and economics', *Journal for Nature Conservation* 14: 135–9.

Bennett, A.F., J.Q. Radford and A. Haslem (2006) 'Properties of land mosaics: Implications for nature conservation in agricultural environments', *Biological Conservation* 133: 250–64.

Blignaut, J., J. Aronson and M. Wit (2014) 'The economics of restoration: Looking back and leaping forward', *Annals of the New York Academy of Sciences* 1332: 34–47.

Bullock, J.M., J. Aronson, A.C. Newton, R.F. Pywell and J.M. Rey-Benayas (2011) 'Restoration of ecosystem services and biodiversity: Conflicts and opportunities', *Trends in Ecology and Evolution* 26: 541–9.

Donald, P.F. and A.D. Evans (2006) 'Habitat connectivity and matrix restoration: The wider implications of agri-environment schemes', *Ecology* 43: 209–18.

Dorrough, J., P.A. Vesk and J. Moll (2008) 'Integrating ecological uncertainty and farm-scale economics when planning restoration', *Journal of Applied Ecology* 45: 288–95.

Greiner, R. and D. Gregg (2011) 'Farmers' intrinsic motivations, barriers to the adoption of conservation practices and effectiveness of policy instruments: Empirical evidence from northern Australia', *Land Use Policy* 28: 257–65.

Haila, Y., K. Henle, E. Apostolopoulou, et al. (2014) 'Confronting and coping with uncertainty in biodiversity research and praxis', *Nature Conservation* 8: 45–75.

Hobbs, R.J. and V.A. Cramer (2008) 'Restoration ecology: Interventionist approaches for restoring and maintaining ecosystem function in the face of rapid environmental change', *Annual Review of Environment and Resources* 33: 39–61.

Jellinek, S., K.M. Parris, D.A. Driscoll and P.D. Dwyer (2013) 'Are incentive programs working?: Landowner attitudes to ecological restoration of agricultural landscapes', *Journal of Environmental Management* 127: 69–76.

Lindenmayer, D.B., A.D. Manning, P.L. Smith, et al. (2002) 'The focal-species approach and landscape restoration: A critique', *Conservation Biology* 16: 338–45.

Manning, A.D., J. Fischer and D.B. Lindenmayer (2006) 'Scattered trees are keystone structures: Implications for conservation', *Biological Conservation* 132: 311–321.

Maron, M., R.J. Hobbs, A. Moilanen, et al. (2012) 'Faustian bargains?: Restoration realities in the context of biodiversity offset policies', *Biological Conservation* 155: 141–8.

McBride, M.F., K.A. Wilson, J. Burger, et al. (2010) 'Mathematical problem definition for ecological restoration planning', *Ecological Modelling* 221: 2243–50.

Michael, D.R., J.T. Wood, M. Crane, R. Montague-Drake and D.B. Lindenmayer (2014) 'How effective are agri-environment schemes for protecting and improving herpetofaunal diversity in Australian endangered woodland ecosystems?', *Journal of Applied Ecology* 51: 494–504.

Mitani, Y., Y. Shoji, and K. Kuriyama (2008) 'Estimating economic values of vegetation restoration with choice experiments: A case study of an endangered species in Lake Kasumigaura, Japan', *Landscape and Ecological Engineering* 4: 103–13.

Pannell, D.J. (2008) 'Public benefits, private benefits, and policy mechanism choice for land-use change for environmental benefits', *Land Economics* 84: 225–40.

Pannell, D.J. and R. Wilkinson (2009) 'Policy mechanism choice for environmental management by non-commercial "lifestyle" rural landholders', *Ecological Economics* 68: 2679–87.

Polasky, S., E. Nelson, E. Lonsdorf, P. Fackler and A. Starfield (2005) 'Conserving species in a working landscape: Land use with biological and economic objectives', *Ecological Applications* 15: 1387–401.

Polyakov, M., D.J. Pannell, R. Pandit, S. Tapsuwan and G. Park (2015) 'Capitalized amenity value of native vegetation in a multifunctional rural landscape', *American Journal of Agricultural Economics* 97: 299–314.

Raymond, C.M. and G. Brown (2011) 'Assessing conservation opportunity on private land: Socio-economic, behavioral, and spatial dimensions', *Journal of Environmental Management* 92: 2513–23.

Robbins, A.S. and J.M. Daniels (2012) 'Restoration and economics: A union waiting to happen?' *Restoration Ecology* 20: 10–17.

Rogers, A.A. (2013) 'Public and expert preference divergence: Evidence from a choice experiment of marine reserves in Australia', *Land Economics* 89: 346–70.

Schaich, H. (2009) 'Local residents' perceptions of floodplain restoration measures in Luxembourg's Syr Valley', *Landscape and Urban Planning* 93: 20–30.

Schultz, E.T., R.J. Johnston, K. Segerson and E.Y. Besedin (2012) 'Integrating ecology and economics for restoration: Using ecological indicators in valuation of ecosystem services', *Restoration Ecology* 20: 304–10.

Shindler, B., R. Gordon, M.W. Brunson and C. Olsen (2011) 'Public perceptions of sagebrush ecosystem management in the Great Basin', *Rangeland Ecology and Management* 64: 335–43.

Thomson, J.R., A.J. Moilanen, P.A. Vesk, A.F. Bennett and R. MacNally (2009) 'Where and when to revegetate: A quantitative method for scheduling landscape reconstruction', *Ecological Applications* 19: 817–28.

Tversky, A. and D. Kahneman (1992) 'Advances in prospect theory: Cumulative representation of uncertainty', *Journal of Risk and Uncertainty* 5: 297–323.

Welsch, J., B. Case and H. Bigsby (2014) 'Trees on farms: Investigating and mapping woody re-vegetation potential in an intensely-farmed agricultural landscape', *Agriculture, Ecosystems and Environment* 183: 93–102.

Wilson, K.A., M. Lulow, J. Burger and M.F. McBride (2012) 'The economics of restoration', *Forest Landscape Restoration* (eds L. David, M. Palle and S. John), Springer, New York, pp. 215–31.

Yoshioka, A., M. Akasaka and T. Kadoya (2014) 'Spatial prioritization for biodiversity restoration: A simple framework referencing past species distributions', *Restoration Ecology* 22: 185–95.

11

Scaling the benefits of agri-environment schemes for biodiversity

Geoffrey Kay

Key lessons

- Agri-environment schemes have mixed outcomes for biodiversity, and more monitoring is needed particularly for certain taxonomic groups.

- Agri-environment scheme effectiveness is heavily reliant on the spatial scale of implementation, and addressing this at local and landscape scales is critical for advancing their application for conservation of biodiversity in agricultural landscapes.

- At local scales, information about how species respond to environmental features, as well as the impact of management actions, could improve site selection and effectiveness of management prescriptions.

- At landscape scales, the offsite benefits of agri-environment schemes could be enhanced by better understanding the impact of surrounding landscape context.

- Incorporating information about the patterns of diversity over large areas, as well as the role and sensitivity of habitat metrics to biodiversity, could greatly enhance the biodiversity benefits of agri-environment schemes.

Demand for agri-environment schemes to counteract global biodiversity loss has resulted in the development of some large, continental-scale agri-environment schemes. While some schemes have been successful in addressing the social and policy elements of farmland conservation (Zammit 2013), very few have been able to demonstrate effective biodiversity outcomes across the scale of program implementation (Whittingham et al. 2007). One of the key reasons for this is that, in order to work across large spatial scales, programs have tended to employ rigid management actions, or a one-size-fits-all approach (Batáry et al. 2011; Kleijn and Sutherland 2003). However, recent studies have demonstrated that the effectiveness of agri-environment schemes is influenced by a number of scale-dependant factors, including the amount invested in agri-environment schemes (Dallimer et al. 2010; Hiron et al. 2013), surrounding landscape context (Batáry et al. 2011; Concepción et al. 2012; Gabriel et al. 2010), and the underlying delivery mechanisms used in scheme design (Hajkowicz et al. 2009; Siriwardena 2010). Designing better agri-environment schemes requires a greater understanding and incorporation of these scale-effects.

One of the critical aspects of scale relates to management rules applied to achieve agri-environment scheme goals (see Figure 11.1). Management rules can be applied at one of two scales: locally at the site (i.e. within a single management unit), or across the whole landscape (i.e. at multiple management units). Irrespective of the goal of a particular agri-environment scheme, both the local and landscape-wide management rules are important for achieving conservation outcomes (Gonthier et al. 2014). If we want to conserve targeted species then it is important to not only protect key habitats but also the potential processes aiding their dispersal and other important aspects of their biology (see metapopulation theory). Conversely, if we want to conserve whole communities, we need to understand how they respond to local-scale management. Despite this recognition, our knowledge and integration of these scale-effects into agri-environment schemes remains very limited (Siriwardena 2010). For example, site-level management actions (such as prescribed or rotational grazing) remain poorly resolved (Briske et al. 2011), and landscape-scale dispersal information is poorly understood for many taxa (Driscoll et al. 2014). Better knowledge of local- and landscape-scale factors that influence conservation outcomes would therefore address a key knowledge gap and provide an opportunity to enhance biodiversity outcomes of agri-environment schemes.

In this chapter, we address this knowledge gap by revealing opportunities to integrate scale-effects to improve agri-environment scheme effectiveness for biodiversity across both local and landscape scales. First, we address the need for biological monitoring over ecologically relevant time frames for quantifying scale-effects on agri-environment schemes. Summarising current knowledge of how local- and landscape-scale factors influence agri-environment schemes, we then provide novel research priorities at these scales.

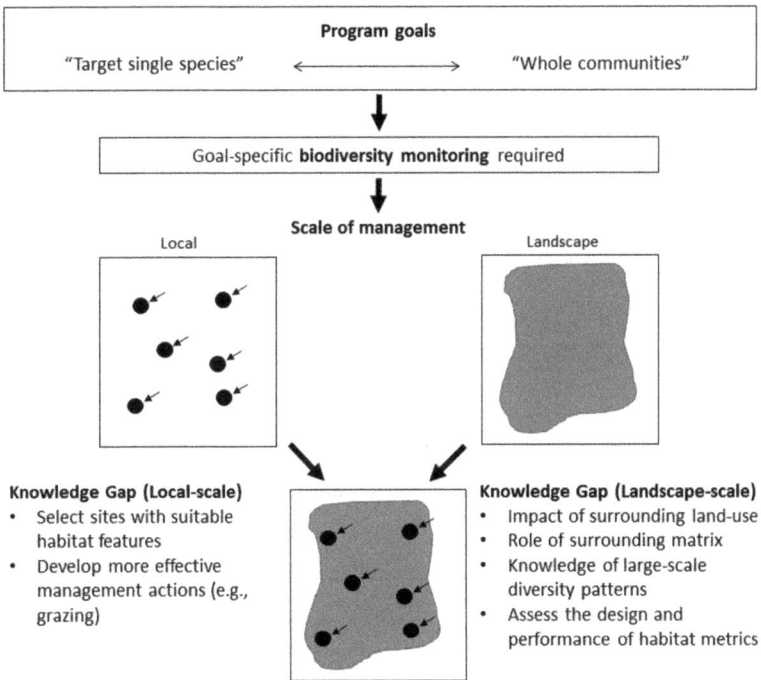

Figure 11.1: Conceptual flow for the advancement of agri-environment schemes through incorporation of scale-effects. Regardless of agri-environment scheme program goals, biodiversity monitoring is required to validate program success. Opportunities to enhance performance of agri-environment schemes by addressing key knowledge gaps at both local- and landscape-scale are identified.
Source: Author's research.

Monitoring outcomes of agri-environment schemes

Quantifying scale-effects on agri-environment schemes relies on a thorough understanding of the biodiversity response to management decisions at different scales. However, despite the significant investment and widespread implementation, many agri-environment schemes have not demonstrated effective outcomes for biodiversity (Kleijn and Sutherland 2003; Lindenmayer et al. 2012; Michael et al. 2014). Amongst other reasons, this is due to a paucity of rigorous assessment and monitoring (Herzog 2005; Perkins et al. 2013), especially for certain taxonomic groups, such as reptiles (Michael et al. 2014). Even where monitoring data is available, our understanding of the scale-effects on biodiversity have largely emerged from what we know of agri-environment schemes in a limited set of regions (e.g. American and European landscapes) (Batáry et al. 2015). To properly assess the biodiversity responses to agri-environment schemes we need to develop robust, statistically verified scientific monitoring programs (Lindenmayer et al. 2012) across a wider set of ecological systems (see Chapter 3). Such programs must be designed around specific agri-environment scheme goals (i.e. single species versus whole communities), and focus on observing population- or community-level changes across multiple taxa at target sites, as well as comparable reference sites.

It is important to consider the time frames necessary for biodiversity response to agri-environment schemes when developing monitoring programs. The inability of many existing monitoring programs to show effective biodiversity response may not reflect failure of the program per se, but that insufficient time has passed for relevant ecological processes to succeed. This emphasises the need to continue monitoring over long time frames far exceeding policy cycles. Critically, however, this does not mean we need to wait until we detect positive change to identify the perfect program. Instead, the limited capacity demonstrated in agri-environment schemes for achieving mid-term outcomes could be addressed by identifying opportunities to enhance the current models.

Advancing agri-environment schemes through understanding scale-effects

Local scale

Fundamentally, the success of any agri-environment scheme relies on the positive response of biodiversity to investment at the local (i.e. field or farm) scale. This is the smallest management scale within an agri-environment scheme, typically 1–10 kilometres, to which landholders apply the funded conservation management actions.

These actions are generally targeted towards specific groups of organisms or agricultural practices, which often include (but are not limited to) promotion of native vegetation, soil health and habitat components (possibly for target species), and prevention of damaging management practices, such as fertilisation and overgrazing (Zammit et al. 2010). Positive biodiversity response in agri-environment schemes depends on two fundamental assumptions: suitable habitat is incorporated within the investment sites selected; and the imposed management actions enhance or maintain suitable conditions (e.g. habitat) for biota. Despite the wide application of agri-environment schemes, major knowledge gaps surround both of these assumptions (see Figure 11.1).

The incorporation of suitable habitat (or the provisioning for future habitat) is critical for achieving biodiversity gain in any agri-environment schemes program. This requires careful consideration of the sites being selected. However, one of the major assumptions regarding site selection is that all habitat within a given ecosystem or species range are equal in condition and extent. For example, native vegetation cover is considered important for a wide number of species (McElhinny et al. 2006) and so is included in the site selection of many Australian agri-environment schemes. However, in many ecosystems targeted under agri-environment schemes, little is known about which habitat features are most important for species. Moreover, the positive influence of specific environmental features is likely to vary for different species, whole communities, and across different biogeographic or climatic zones (Whittingham et al. 2007). It may be important to ask whether targeting certain habitat features, such as those that are important for rare species, can improve the effectiveness of agri-environment schemes for other biota. Indeed, some of the most important drivers of biodiversity may be climatic or landscape features (e.g. elevation), which cannot be influenced by management.

Central to agri-environment schemes policy is the use of carefully prescribed site-level management actions, which counteract the negative influence of agriculture on biodiversity. Management actions addressing the impact on biodiversity from wide-spread agricultural practices (e.g. livestock overgrazing, fertiliser application) are considered most desirable (Dallimer et al. 2010). Despite a focus on these management actions in agri-environment schemes, their role in averting biodiversity loss is poorly known. For example, there is little consensus on the impacts of livestock grazing as a management tool for biodiversity, despite widespread application and decades of research (Briske et al. 2011). A better understanding of the outcomes of management actions on biodiversity (particularly large-scale ones, such as livestock grazing) would have large implications for enhancing site-level response to agri-environment schemes.

Landscape scale

An underlying basis for successful agri-environment schemes is that they propagate positive biodiversity benefit from investment sites into the broader landscape. The extent to which this occurs depends heavily on the hostility of the surrounding landscape (Tscharntke et al. 2005). Yet the mechanisms remain poorly understood. This is particularly the case for species with limited dispersal (e.g. terrestrial invertebrates, reptiles), which are most sensitive to the negative impacts of fragmentation in an agricultural matrix.

Recent studies have found that the biodiversity response to agri-environment scheme investment is influenced by the context of the surrounding landscape (Carvell et al. 2011; Concepción et al. 2012), although this response is inconsistent and poorly resolved (Batáry et al. 2011). The greatest positive biodiversity response has been observed in landscapes with an intermediate level of 'complexity' — the degree of variation within landscape (Concepción et al. 2012). Despite this, other studies have found inconsistent results for different taxonomic groups for both simple and complex landscapes, and that the effect of complexity depends on the type of surrounding vegetation (Batáry et al. 2011). Landscapes with a greater proportion of area covered by management demonstrate stronger positive biodiversity response (Baker et al. 2012; Dallimer et al. 2010), although whether this is in response to direct agri-environment scheme investment or greater inclusion of suitable landscapes remains unclear. A greater number of studies exploring the influence of surrounding landscape on agri-environment schemes success are clearly required to systematically investigate each of these conflicting elements. Given that our current understanding is nearly exclusively from European (e.g. Concepción et al. 2012) and American landscapes (e.g. Carvell et al. 2011), future investigations of this kind within Australian landscapes are critical for providing a more rounded understanding of how to enhance agri-environment schemes at landscape scales.

In addition to understanding how landscape context can affect agri-environment schemes success, it is important to know why landscape context may affect agri-environment schemes success. One of the key assumptions in agri-environment schemes policy is that investment will promote even propagation of biodiversity into the broader landscape (Whittingham et al. 2007), although it is clear that dispersal into surrounding landscapes for some species (e.g. ground-dependant species such as reptiles) will be more challenging than others. This is because the type of land-use and style of management employed in surrounding landscapes is likely to greatly influence the capacity of species to disperse, particularly for those with limited dispersal ability most at risk to fragmentation. For agri-environment schemes to better support the conservation of such species in the broader landscape, it is crucial that we gain a greater understanding of how these species disperse in different matrix environments. Non-hostile matrix environments can then be created. We can achieve this by examining the effect of surrounding land-use (e.g. cropped landscapes), as well as the impact of certain management actions (e.g. grazed and open pastures versus ungrazed and dense pastures), on the movement of limited-dispersing species.

The capacity for agri-environment schemes to achieve landscape-scale, positive biodiversity response is dependent largely on the ability of these schemes to adequately capture the best elements of the landscape. Agri-environment schemes that pay landholders for biodiversity actions create a market for biodiversity, but the value assigned to actions is determined by metrics adopted in the scheme (Zammit et al. 2010). Despite widespread use of habitat metrics in biodiversity markets, relatively little attention has been paid to their design and performance (Hajkowicz et al. 2009), and their capacity to summarise the actual biodiversity present at a site within an agri-environment scheme (Oliver et al. 2014). The use of metrics that misrepresent the condition of the target ecological community has dire implications for the overall success of an agri-environment scheme, and yet we have very little knowledge linking biodiversity response to metric scores for agri-environment schemes worldwide.

Finally, landscape-wide biodiversity outcomes for agri-environment schemes can also be achieved by developing a broader understanding of the patterns of diversity across the landscape. Currently, the distribution of diversity is known at a very general level — for example, species richness is greater near the equator (Gaston 2000) — with little or no understanding of diversity patterns at landscape scales for most groups of organisms. Understanding landscape patterns of diversity would have considerable benefit for agri-environment schemes by allowing managers to develop regionally targeted conservation strategies. For example, focused management of low-quality sites within high diversity areas (hotspots) could be used to achieve high conservation gain. Although the tools for developing such diversity maps have now been developed (Ferrier et al. 2007), they have yet to be integrated into agri-environment schemes, despite the clear conservation benefits of doing so.

Conclusions

Enhancing the effectiveness of agri-environment schemes for biodiversity conservation requires management attention at multiple scales, from local to landscape. We have identified key knowledge gaps and priority areas for research that would improve the rigid one-size-fits-all model commonly applied to agri-environment schemes. We contend that the effectiveness of agri-environment schemes must be advanced if we are to counter the effects of agriculture on biodiversity, and that monitoring data across multiple scales, for a much wider range of taxa, is required.

References

Baker, D.J., S.N. Freeman, P.V. Grice, et al. (2012) 'Landscape-scale responses of birds to agri-environment management: A test of the English Environmental Stewardship scheme', *Ecological Applications* 49: 871–82. DOI:10.1111/j.1365-2664.2012.02161.x.

Batáry, P., A. Báldi, D. Kleijn and T. Tscharntke (2011) 'Landscape-moderated biodiversity effects of agri-environmental management: A meta-analysis', *Proceedings of the Royal Society B: Biological Sciences* 278: 1894–902. DOI:10.1098/rspb.2010.1923.

Batáry, P., L.V. Dicks, D. Kleijn and W.J. Sutherland (2015) 'The role of agri-environment schemes in conservation and environmental management', *Conservation Biology* 29(4): 1006–16. DOI:10.1111/cobi.12536.

Briske, D.D., N.F. Sayre, L. Huntsinger, et al. (2011) 'Origin, persistence, and resolution of the rotational grazing debate: Integrating human dimensions into rangeland research', *Rangeland Ecology and Management* 64: 325–34. DOI:10.2111/REM-D-10-00084.1.

Carvell, C., J.L. Osborne, A.F.G. Bourke, et al. (2011) 'Bumble bee species' responses to a targeted conservation measure depend on landscape context and habitat quality', *Ecological Applications* 21: 1760–71. DOI:10.1890/10-0677.1.

Concepción, E.D., M. Díaz, D. Kleijn, D., et al. (2012) 'Interactive effects of landscape context constrain the effectiveness of local agri-environmental management', *Ecological Applications* 49:, 695–705. DOI:10.1111/j.1365-2664.2012.02131.x.

Dallimer, M., K.J. Gaston, A.M.J. Skinner, et al. (2010) 'Field-level bird abundances are enhanced by landscape-scale agri-environment scheme uptake', *Biology Letters* 6: 643–6. DOI:10.1098/rsbl.2010.0228.

Driscoll, D.A., S.C. Banks, P.S. Barton, et al. (2014) 'The trajectory of dispersal research in conservation biology: Systematic review', *PLoS ONE* 9: e95053. DOI:10.1371/journal.pone.0095053.

Ferrier, S., G. Manion, J. Elith and K. Richardson (2007) 'Using generalized dissimilarity modelling to analyse and predict patterns of beta diversity in regional biodiversity assessment', *Diversity and Distributions* 13: 252–64. DOI:10.1111/j.1472-4642.2007.00341.x.

Gabriel, D., S.M. Sait, J.A. Hodgson, et al. (2010) 'Scale matters: the impact of organic farming on biodiversity at different spatial scales', *Ecology Letters* 13: 858–69. DOI:10.1111/j.1461-0248.2010.01481.x.

Gaston, K.J. (2000) 'Global patterns in biodiversity', *Nature* 405: 220–27. DOI:10.1038/35012228.

Gonthier, D.J., K.K. Ennis, S. Farinas, et al. (2014) 'Biodiversity conservation in agriculture requires a multi-scale approach', *Proceedings of the Royal Society B: Biological Sciences* 281: 20141358. DOI:10.1098/rspb.2014.1358.

Hajkowicz, S., K. Collins and A. Cattaneo (2009) 'Review of agri-environment indexes and stewardship payments', *Environmental Management* 43: 221–36. DOI:10.1007/s00267-008-9170-y.

Herzog, F. (2005) 'Agri-environment schemes as landscape experiments', *Agriculture, Ecosystems and Environment* 108: 175–177. DOI:10.1016/j.agee.2005.02.001.

Hiron, M., Å. Berg, S. Eggers, et al. (2013) 'Bird diversity relates to agri-environment schemes at local and landscape level in intensive farmland', *Agriculture, Ecosystems and Environment* 176: 9–16. DOI:10.1016/j.agee.2013.05.013.

Kleijn, D. and W.W. Sutherland (2003) 'How effective are European agri-environment schemes in conserving and promoting biodiversity?', *Journal of Applied Ecology* 40, 947–69.

Lindenmayer, D.B., C. Zammit, S.J. Attwood, et al. (2012) 'A novel and cost-effective monitoring approach for outcomes in an Australian biodiversity conservation incentive program', *PLoS ONE* 7: 1–11. DOI:10.1371/journal.pone.0050872.

McElhinny, C., P. Gibbons, C. Brack and J. Bauhus (2006) 'Fauna-habitat relationships: A basis for identifying key stand structural attributes in temperate Australian eucalypt forests and woodlands', *Pacific Conservation Biology* 12: 89–110.

Michael, D.R., J.T. Wood, M. Crane, R. Montague-Drake and D.B. Lindenmayer (2014) 'How effective are agri-environment schemes for protecting and improving herpetofaunal diversity in Australian endangered woodland ecosystems?' *Journal of Applied Ecology* 51: 494–504. DOI:10.1111/1365-2664.12215.

Oliver, I., D.J. Eldridge, C. Nadolny and W.K. Martin (2014) 'What do site condition multi-metrics tell us about species biodiversity?' *Ecological Indicators* 38: 262–71. DOI:10.1016/j.ecolind.2013.11.018.

Perkins, A., H. Maggs, A. Watson and J. Wilson (2010) 'Adaptive management and targeting of agri-environment schemes does benefit biodiversity: A case study of the corn bunting Emberiza calandra', *Journal of Applied Ecology* 48: 514–22.

Perkins, G., A. Kutt, E.P. Vanderduys, J. Perry and J.J. Perryl (2013) 'Evaluating the costs and sampling adequacy of a vertebrate monitoring program', *Australian Journal of Zoology* 36: 373–80.

Siriwardena, G.M. (2010) 'The importance of spatial and temporal scale for agri-environment scheme delivery', *Ibis* 152(3): 515–29. DOI:10.1111/j.1474-919X.2010.01034.x.

Tscharntke, T., A.M. Klein, A. Kruess, I. Steffan-Dewenter and C. Thies (2005) 'Landscape perspectives on agricultural intensification and biodiversity: Ecosystem service management', *Ecology Letters* 8, 857–74. DOI:10.1111/j.1461-0248.2005.00782.x.

Whittingham, M.J., J.R. Krebs, R.D. Swetnam, et al. (2007) 'Should conservation strategies consider spatial generality?: Farmland birds show regional not national patterns of habitat association', *Ecology Letters* 10: 25–35. DOI:10.1111/j.1461-0248.2006.00992.x.

Zammit, C., S. Attwood and E. Burns (2010) 'Using markets for woodland conservation: Lessons from the policy-research interface', *Temperate Woodland Conservation and Management* (eds D.B. Lindenmayer, A.F. Bennett and R.J. Hobbs), CSIRO Publishing, Melbourne, pp. 297–307.

Zammit, C.C. (2013) 'Landowners and conservation markets: Social benefits from two Australian government programs', *Land Use Policy* 31: 11–16. DOI:10.1016/j.landusepol.2012.01.011.

12

Social dimensions of biodiversity conservation programs

Saan Ecker

Key lessons

- Respond to landholder motivations for biodiversity conservation.
- Understand the socio-demographic profile of potential participants.
- Support those already making the change.
- Design programs to achieve compatibility between financial and biodiversity outcomes.

Social dimensions play an important role in landholder participation in natural resource management (NRM) programs. Many regional- and community-based NRM organisations have a good understanding of landholder characteristics and capacity from data collected through national, regional, or catchment scale landholder surveys and other social studies. But, often, NRM plans only include aspirations for integrating this data into program design, project communications and evaluation. While various frameworks have been developed to do this (Fenton 2004; Robins and Dovers 2007), there are few recorded cases where integration of social, environmental and economic information has been used to support successful implementation of NRM programs (Bammer et al. 2005).

This chapter draws on two examples in which landholder characteristics were examined to gain insight into how specific NRM programs could be better implemented. The case studies used in this chapter are former Australian Government NRM programs: the Environmental Stewardship Program, and the Sustainable Farm Practice component of the Caring for our Country (CofC) initiative. A study into the motivations, drivers, and barriers associated with involvement in the Environmental Stewardship Program was undertaken at the beginning of the program. A longitudinal national survey on adoption and reasons for adoption was undertaken at the beginning and end of the Sustainable Farm Practice component of the CfoC initiative. Findings from these studies support five social dimensions considered important in the development and implementation of biodiversity conservation programs.

Respond to landholder motivations for biodiversity conservation

Human behaviour is complex. At the very least, decisions to act include the combined influences of behaviours, intentions, perceived behavioural control, attitude and societal norms (Fishbein and Ajzen 2010). Lack of understanding of what drives biodiversity conservation can lead to simplistic policies and potentially alienate possible participants. Effort in understanding landholders' complex motivations for biodiversity conservation is likely to benefit outcomes of biodiversity initiatives, particularly when this understanding is embedded in program design and extension.

For example, the national Drivers of Practice Change survey canvassed 1,400 commercial farmers to identify their motivations for adoption of sustainable land management practices which were promoted under the CfoC initiative (Ecker et al. 2012). The survey asked respondents to select one of three motivational areas (financial, environmental, and personal) for each practice. Farmers rated environmental factors as most frequently influencing their adoption of native vegetation management practices. Respondents selected up to three detailed statements for each motivational area (see Table 12.1). These provided insight into the diverse factors that farmers considered in facilitating adoption of these practices.

Table 12.1: Most frequently selected motivational areas and motives for native vegetation management practices, listed in order of importance to respondents.

	Motives		
	Financial benefits	Environmental factors	Personal motivations
Motivational area	Provides shelter for livestock Increased land value Increased returns/ income	Improves soil quality Aligns with environmental goals and beliefs Provides habitat for fauna	Desire to protect natural resources Desire to improve amenity of the landscape Recognition by neighbours and community

Source: Ecker et al. (2012); Kancans et al. (2014).

The results from a different survey on landholder motivations for involvement in the Environmental Stewardship Program also demonstrated complex considerations influencing participation in conservation activities (Ecker and Thompson 2011). Financial motivations, environmental stewardship interests, prior conservation efforts, the opportunity to engage in a social network, and the opportunity to learn more about native vegetation management contributed to landholder participation in the program. Most respondents 'strongly agreed' that conservation and enhancement of native vegetation contributed to improved property or landscape health, aesthetics, soil stabilisation, and controlling rising water tables. Few thought that conservation and enhancement of native vegetation increased fire risk, and fewer saw it as an impediment to controlling pests and weeds. Understanding what actually motivates people to participate in biodiversity initiatives can assist with engagement strategies and lead to better communication and relationships with participants.

Understand the socio-demographic profile of potential participants

The farm business and farmer's personal characteristics are important to consider in program design of and stakeholder engagement with biodiversity protection initiatives. Social profiling has been widely used by regional NRM groups to better understand the socio-demographic profile of communities. However, the usefulness of this information depends on scale and the quality of data collection and its interpretation.

In the context of the CfoC initiative, Kancans et al. (2014) explored demographic and other adoption-related characteristics relevant to land management practices. Table 12.2 reflects the findings from a series of one-way analysis of variance, comparing the characteristics of adopters and non-adopters of each practice. These findings show that landholders more likely to conserve or manage native vegetation are older, well-educated, have a strong financial status, are a member of a land management group, and have prior involvement in a government land management initiative.

Table 12.2: Characteristics more likely to be found in adopters of specific land management practices.

Adopter characteristic	Land management practice		
	Native pasture conservation or management	Native vegetation conservation or management	Fencing native vegetation
Higher cash income	No	Yes	Yes
Higher rate of return	No	Yes	Yes
Larger farms	Yes	No	No
Younger	No	No	No
Higher level of education	Yes	Yes	Yes
Participate in government program	Yes	Yes	Yes
Participate in extension	No	No	No
Member of land management group	Yes	Yes	Yes

Source: Kancans et al. (2014)

A landholder's previous experience with an NRM program has been shown to influence their involvement in a future conservation initiative (Windle and Rolfe 2006). Supporting this, Ecker and Thompson (2011) found that 45 per cent of applicants in the Environmental Stewardship Program had previously been involved in environmental or NRM programs, including financial and non-financial support.

Farm income is often cited as an import driver of land management practice adoption, however, the influence of farm income on conservation adoption is ambiguous, as financial assistance from the NRM program may overcome farm income constraints (Cary et al. 2002). Another important aspect to understand is baseline knowledge in regard to conservation practices. Seventy per cent of Environmental Stewardship Program applicants said they had never undertaken training relevant to native vegetation conservation, and less than half could identify box gum grassy woodlands on their property (Ecker and Thompson 2011). Baseline information such as this can be useful for assessing enduring change in the long term.

Support those already making the change

There is evidence that external support sources are secondary to intrinsic motivations in NRM decisions. For example, Farmar-Bowers and Lane (2006) suggest that farmers use a number of lenses when seeking and identifying an opportunity, starting with personal motivations related to the opportunity (intrinsic interests, family considerations, and personal knowledge), before moving onto external components, such as knowledge and support. It is likely that support is associated with increasing capacity to implement new management practices once landholders have decided to adopt. This is the point at which they seek support. The implication is that support is better received by those who have made the decision to implement the practice. As obvious as it sounds, this thinking is not always embedded in policy and program approaches.

Lending some support to this concept, Ecker and Thompson (2011) found that the majority of participants applying for the Environmental Stewardship Program had previously participated in conservation activities, with more than 80 per cent saying they had revegetated parts of their property, and more than half having fenced remnant vegetation. Repeat customers may be frowned upon in program evaluations, as there is often a preference within the program to attract new participants. Targeting landholders who are ready and able to make changes, and support the improvement of biodiversity values over time, may be a more realistic and efficient target. This is supported by Greiner and Gregg (2011), who demonstrated that landholders with higher intrinsic motivation (i.e. higher scores for 'lifestyle and stewardship motivation') had undertaken more conservation action on their properties.

Ecker et al. (2012) outlined the role that support providers play in influencing landholder decisions to adopt land management practices. Non-financial support played a secondary role to financial, environmental, and personal motivations in decisions to adopt NRM activities. Relatively few respondents (9 per cent) said the availability of non-financial support influenced them in land management practice decisions 'to a great extent'. While support is of critical importance in maintaining the impetus for adoption of native vegetation management, these findings suggest that support is generally accessed after the decision to adopt the practice is made.

Recognise the importance of community-based conservation and NRM organisations

Community-based groups, such as regional NRM groups (often a mix of government and community), Landcare, catchment, conservation and other care groups, maintain momentum and continuity between shifts in government policies. As noted in Chapter 5, environmental NGOs are often more focused on on-ground activities and, as such, are in the best position to facilitate between individuals and national environmental challenges. When the groups adequately represent social catchments, they provide foci and forums for NRM in the community. While the importance of community-based groups is well covered in the literature (e.g. Marshall 2010), quantitative measures that provide substantive evidence of the importance of these groups is harder to come by.

Results of the national Drivers of Practice Change survey on farmer motivations for sustainable land management practices, administered in 2010 and 2012, demonstrated the importance of community groups in supporting landholders' adoption of native vegetation management practices from three different survey questions. First, Landcare or farmer production groups were found to be the most important influence on native vegetation management adoption (Ecker et al.

2012; Kancans et al. 2014). Second, of the 27 per cent of adopters who sought non-financial support, over half said they obtained this support from Landcare groups, followed by catchment management authorities (CMAs) (20 per cent). Third, as shown in Table 12.2, members of NRM groups, including community- and production-focused groups, were more likely to adopt native vegetation management practices than non-members (Kancans et al. 2014).

Ecker and Thompson (2011) also found that around 60 per cent of participants involved in the Environmental Stewardship Program indicated that CMAs are an important information source when making decisions to conserve native vegetation. They considered CMAs more important than any other information source. The importance of Landcare as a source of information to participants was also evident: 44 per cent of the participants indicated that Landcare groups were important in their decision-making processes related to native vegetation management. The influence of these community-based groups is moderated by factors such as longevity and degree of connection with the community.

A relationship with non-government NRM and conservation organisations, through both regular (e.g. extension) and irregular (e.g. forums and events) interactions, and the degree of trust held and the valuing of this support provision is known to affect landholders involvement in land management programs (Jennings 2005). Landcare groups and other community-based organisations, such as regional NRM groups and CMAs, established in the community are well positioned to deliver or partner with other delivery agents in biodiversity program design and implementation, and have an important role to play in the long-term success of these initiatives.

Design programs to achieve compatibility between financial and biodiversity outcomes

A profitable farming enterprise is an important goal for farmers. It is important that biodiversity conservation initiatives consider this goal in program design and extension. The phrase 'need to be in the black to be in the green' is popular amongst farmers. The first strategy in this chapter emphasised the interrelatedness of financial, environmental, and personal motivations. While the environmental stewardship motivations of those involved in biodiversity conservation are clear, a balance between environmental and financial outcomes is important for some farmers.

Lockie and Tennent (2010) note issues with previous schemes included an inability to coherently link farm operations and biodiversity outcomes. In the study by Ecker and Thompson (2011), participants said that the Environmental Stewardship Program had succeeded in linking production farming and biodiversity outcomes through the flexibility allowed for productive use of conservation areas under certain circumstances. Landholders who rely on farm income need to have adequate flexibility to manage drought conditions and other tough financial periods. Participants in the study were generally well informed about the impacts and benefits of different grazing regimes on both biodiversity outcomes and profit, including understanding the importance of having functioning grassy woodland ecosystems in managing livestock through these tough periods. Many participants indicated that they had reduced stocking numbers in order to maintain native pastures, and that in some cases this had improved the quality of wool and lambing percentages. Encouraging debate, discussion, and shared learning on how to best maximise profit and conservation outcomes, and subsequent incorporation of this information into program design is important in ensuring the long-term success of biodiversity initiatives.

Summary

This chapter draws on recent studies on landholder adoption of biodiversity conservation programs to lend support to five key strategies that are important in ensuring the success of such initiatives: respond to landholder motivations for biodiversity conservation; understand the socio-demographic profile of potential participants; support those already making the change; recognise the importance of community-based conservation and NRM organisations; and design programs to achieve compatibility between financial and biodiversity outcomes. These strategies relate to understanding and responding to the complex human ecology of participation in biodiversity initiatives where financial, environmental and personal factors interrelate to enhance or impede this participation. While past NRM programs have adopted some of these strategies, rarely are all these factors considered in program design. Time spent in improving the understanding of the target audience, building a strong community-based support network, recognising past initiatives, and piloting the program prior to broad-scale implementation is time well spent. These five strategies, amongst the other insights recorded in this book, can help towards building effective approaches to achieve successful biodiversity outcomes that benefit both environment and society.

References

Bammer, G., C. Mobbs, R. Lane, S. Dovers and C. Allan (2005) 'An introduction to Australian case studies of integration in natural resource management', *Australasian Journal of Environmental Management* 12 (supplementary issue).

Cary, J., T. Webb and N. Barr (2002) 'Understanding land managers' capacity to change to sustainable practices: Insights about practice adoption and social capacity for change', Bureau of Rural Sciences, Canberra.

Commonwealth of Australia (2002) *National natural resource management capacity building framework*, Department of Agriculture, Fisheries and Forestry and Environment and Heritage, Canberra.

Cullen, P. (2004) 'Using local and scientific knowledge to inform catchment management', presentation to Upper Murrumbidgee Catchment Coordinating Committee Biennial Natural Resource Management Forum, 29 September, University House, The Australian National University, Canberra.

Ecker, S. and L.J. Thompson (2011) *Participation in the Environmental Stewardship Program Box Gum Grassy Woodlands Project: key findings and implications*, ABARES report to client for the Department of Sustainability, Environment, Water, Population and Communities, Canberra.

Ecker, S., L. Thompson, R. Kancans, N. Stenekes, and T. Mallawaarachchi (2012) 'Drivers of practice change in land management in Australian agriculture', ABARES, Department of Agriculture, Fisheries and Forestry, Canberra.

Farmar-Bowers, Q. and R. Lane (2006) *Understanding farmer decision systems that relate to land use*, Department of Sustainability and Environment, Victoria, report to the School of Global Studies, Social Sciences and Planning, RMIT, Melbourne.

Fenton, D.M. (2004) *A monitoring and evaluation framework for the social dimensions of the NHT and NAPSWQ*, National Land and Water Resource Audit, Canberra.

Fishbein, M. and I. Ajzen (2010) *Predicting and changing behavior: The reasoned action approach*, New York, Taylor & Francis.

Greiner, R. and D. Gregg (2011) 'Farmers' intrinsic motivations, barriers to the adoption of conservation practices and effectiveness of policy instruments: Empirical evidence from northern Australia', *Land Use Policy* 28: 257–65.

Jennings, J. (2005) *On the Effectiveness of Participatory Research in Agriculture*, PhD thesis, University of Western Sydney.

Kancans, R., S. Ecker, A. Duncan, N. Stenekes and H. Zobel-Zubrzycka (2014) *Drivers of practice change in Australian agriculture: Synthesis report: Stages I, II and III*, Australian Government Department of Agriculture, Canberra.

Lockie, S. and R. Tennent (2010) 'Market instruments and collective obligations for on-farm biodiversity conservation', *Agriculture, Biodiversity and Markets Livelihoods and Agroecology in Comparative Perspective* (eds S. Lockie and D. Carpenter), Earthscan, London.

Marshall, G.R. (2010) 'What "community" means for farmer adoption of conservation practices', *Changing land management: Adoption of new practices by rural landholders* (eds D. Pannell and F. Vanclay), CSIRO Publishing, Melbourne.

Pannell, D.J., G.R. Marshall, N. Barr, A.F.V. Curtis and R. Wilkinson (2006) 'Understanding and promoting adoption of conservation technologies by rural landholders', *Australian Journal of Experimental Agriculture* 46(11): 1407–24.

Robins, L. and S. Dovers (2007) 'NRM regions in Australia: The haves and the have nots', *Geographical Research* 45(3): 273–90.

Rogers, E. (2003) *Diffusion of Innovations*, fifth edition, Simon and Schuster, London.

Windle, J. and J. Rolfe (2006) *Fitzroy Basin Association's Biodiversity Tender: An outline and evaluation*, Central Queensland University, Rockhampton.

13

Contract preferences and psychological determinants of participation in agri-environment schemes

Romy Greiner

Key lessons

- Agri-environment schemes offer positive financial incentives to farmers, and are favoured by farmers over other policy approaches, but this does not translate into unconditional participation. To maximise participation, agri-environment scheme design needs to consider farmer preferences for contract features, motivations, and attitudes.

- In general, farmers are more likely to sign up to agri-environment scheme contracts that allow some form of agricultural production on the contract area, offer a higher per hectare payment, are shorter, allow flexibility, and are externally monitored.

- Preferences are context specific and there is significant variation in preferences among farmers, meaning that a suite of agri-environment scheme contract options works best to maximise participation. But contract choice must not compromise the intended conservation result.

- Agri-environment schemes need to be supported by complementary measures, such as information and extension, to shape attitudes and ensure that agri-environment scheme design, implementation, and administration do not jeopardise existing altruism and intrinsic motivation for conservation among farmers.

- Biodiversity conservation on private land, funded by voluntary contractual arrangements, is likely to be an expensive way to do conservation. From the perspective of efficiency and permanency of investment, inclusion of land into the formal conservation estate is preferable. However, agri-environment schemes can play a vital role of securing strategically important areas into a multi-tenure conservation system in the short- to medium-term.

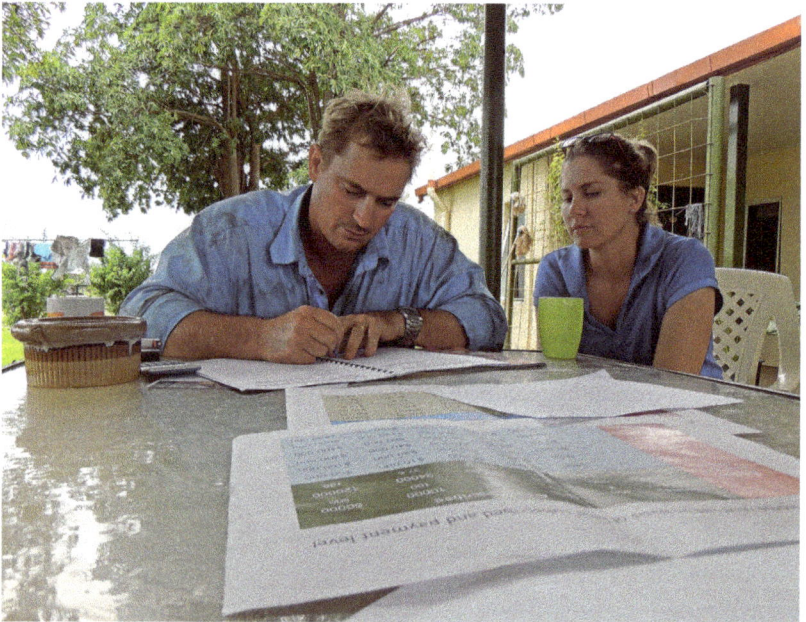

Figure 13.1: Conducting research on pastoralists' participation in contractual biodiversity conservation at Hayfield Station in the Northern Territory.
Source: Photo by Romy Greiner.

Paying farmers for environmental services is a novel concept in the vast landscapes of northern Australia. These landscapes remain sparingly used and the tropical savannas vegetation has been largely retained (Greiner et al. 2009a). But degradation is occurring, and pressures for agricultural development and intensification are mounting. The opportunity still exists to prevent the scale of environmental decline and species extinctions that has been experienced in other parts of Australia. Could agri-environment schemes be an effective policy instrument to safeguard northern Australia's biodiversity, and, if so, what should an effective agri-environment scheme look like?

Europe has 30 years' experience with conservation programs, which provide incentives directly to farmers to protect and manage land for biodiversity. A recent review of these biodiversity agri-environment schemes has shown that they can be effective for conserving wildlife on farmland, but that agri-environmental schemes need to be carefully designed to achieve their goals (Batáry et al. 2015). Schemes need to be feasible across large landscapes and incentivise landholder participation.

An obvious aspect of agri-environmental scheme design is deciding on the level of incentive to pay to participating landholders. Microeconomic theory of profit maximisation might suggest that landholders will consider participating in a scheme if they do not incur a net loss of income. Assuming that a switch to conservation-focused land management causes a loss of income in most circumstances and can result in additional costs, the financial incentive has to at least compensate these imposts. However, a singular focus on the financial dimension is misguided, because adoption and participation decisions by farmers are influenced by consideration of various aspects of the innovation (Pannell et al. 2006). In the context of agri-environment schemes this means that, in addition to its immediate financial dimension, other contract attributes are also important, as are social and psychological factors (Greiner et al. 2009b).

As the theory of planned behaviour explains, psychological factors such as motivations and attitudes can be important predictors of behaviour (Ajzen 1991), including adoption of innovation and program participation. Knowledge of which factors influence participation, and how, can help with the design and implementation of agri-environment schemes so as to maximise farmer participation, improve scheme effectiveness, and maximise efficiency of investment.

This chapter summarises relevant knowledge and offers illustrations based on a survey of 104 farmers in northern Australia, as part of the Australian Government's National Environmental Research Program Northern Australia Hub. Face-to-face interviews were conducted with pastoral land owners and land managers, mostly on their properties. Respondents represent approximately 15 per cent of farmers in the rangelands of the tropical savannas and manage a combined area of over 250,000 km² of land. Methodological specifications and detailed results can be gleaned from Greiner (2014a, 2015) and Greiner et al. (2014).

Figure 13.2: The Einasleigh Uplands are one landscape within northern Australian tropical savannas.
Source: Photo by Romy Greiner.

Farmers have diverse preferences for contract attributes

Agri-environment schemes are principally implemented using contracts between purchasers of conservation services, typically governments, and farmers as providers of various conservation services. In essence, an agri-environment scheme constitutes a payment for environmental services (Greiner et al. 2009a). Contract attributes as a minimum entail the conservation actions that farmers have to undertake, the level of stewardship payment they receive in return ($/ha), and duration of the agreement (years). They also tend to stipulate arrangements about monitoring and administration, and possible flexibility provisions.

The payment levels offered in the choice experiment were guided by long-term estimates of farm cash income in the tropical savannas and industry consultations. Across the tropical savannas, the average annual cash income is approximately $3 per hectare (DAFF 2014, 2013 equivalent values). However, income is highly variable, both temporally and spatially: the coefficient of variation between 1990–2013 is close to 100 per cent, and in good seasons cash income can be four times the average income, while in bad seasons cash income is negative. Across the tropical savannas there are vast areas of marginal land with zero productivity while some pockets of fertile land average 10 times the landscape average cash income.

A body of literature is dedicated to exploring how farmers trade-off contract attributes against each other and the per hectare payment (e.g. Garrod et al. 2012; Burton and Schwarz 2013; Espinosa-Goded et al. 2010; Broch and Vedel 2012). The literature shows that, in general, farmers prefer higher payments, shorter contracts, more flexibility, less accountability, and less paperwork. Preferences and trade-offs are context specific, meaning they are dependent on the geographical setting, type of agricultural sector, and the conservation goal to be achieved.

Observed and stated preference techniques can be used to quantify farmers' preferences and how they trade off between contract attributes, and to explain how each attribute affects the likely participation by farmers in the scheme. Preferences indicate the monetary value that farmers attach to contract attributes. If, for example, a farmer's preference is for a short contract and the funder wants to offer a long contract, then the funder will most likely need to offer a higher per-hectare stewardship payment to secure the farmer's participation.

The research with northern Australian farmers used a stated preference technique — choice experiment — to explore participation in hypothetical biodiversity conservation contracts

(Greiner et al. 2014). Biodiversity in this context was defined as native flora and fauna, and the ecosystems that support them. Analysis of the choice experimental data confirmed the aforementioned generic findings in the literature regarding preferences for contract features and provided context specific preference estimates (Greiner 2015). For example, across the industry, graziers were asking for a $0.40 increase in annual per hectare payment (value in 2013 equivalent) for an additional year of contract duration. In terms of the conservation requirement, across the industry, graziers required an extra $11.08 per hectare a year (95 per cent confidence interval: $7.45–$14.47) as a premium for participation in contracts for the complete removal of cattle from the contract area, compared to contract options that required the exclusion of cattle for only a short period of time each year (such as when the biodiversity was most susceptible to grazing impact). This would suggest that, across the industry, landholders are benchmarking opportunity costs against income in favourable years rather than long-term average conditions — they are hedging their bets with participation in agri-environment schemes.

The results of that choice experiment also illustrated the significant variation in preferences for all contract attributes (Greiner 2015). For example, across the industry, some farmers preferred that monitoring of contract compliance be undertaken by external providers, and others preferred to undertake the monitoring themselves (with occasional spot checks). Some farmers charged a very high premium for total exclusion of cattle, while others were prepared to accept this option more cheaply, all else being equal — but still at a premium above opportunity cost. Understanding the diversity of preferences for contract attributes — even within a seemingly homogenous farming sector such as the northern beef industry — helps investors to gauge the likely responses by farmers in a region or in an industry to certain contract features. This information consequently enables agri-environment scheme design to be tailored — contractually and administratively — so as to ensure fit for purpose and program efficiency.

Making money is important but other goals may be more important to some

Age and education are the personal characteristics of farmers most often associated with participation in agri-environment schemes (see Chapter 10). However, in the research with northern Australian farmers, neither of these characteristics was found to have a significant influence (Greiner 2015). Psychological constructs, such as attitudes and motivations, seem to be more relevant to behaviour. Both have previously been found to be antecedents of conservation behaviour of northern Australian farmers (Greiner and Gregg 2011). Empirical research with farmers elsewhere has similarly established the importance of motivations (see Chapter 12). To test this theory in the context of agri-environment scheme participation, the survey included separate five-point Likert scales to derive measures of motivation and attitudes towards biodiversity (Greiner 2014a).

The research found that the five most highly rated goals (based on mean value of motivation items) did not include financial items but were to: 'look after cattle', 'get satisfaction from living and working on the land', 'pass on land in good condition', 'enjoy life and work on the property', and 'look after the natural assets of the property' (see Table 13.1). Survey results supported the idea that farmers in northern Australia have a high intrinsic stewardship motivation for safeguarding their cattle, land, and biodiversity assets, and that this is fundamentally linked to the pursuit of pastoralism as a chosen lifestyle (Greiner and Gregg 2011).

Principal component factor analysis was used to group the goals into broad themes, or factors, which represented three different types of motivations: 'stewardship and lifestyle motivation', 'social motivation', and 'economic/financial motivation'. Respondents who tended to have high ratings for stewardship and lifestyle motivation were driven by a custodianship ethic ('look after the natural assets', 'look after cattle') combined with enjoyment of their work and lifestyle. Farmers who tended to score highly on economic/financial motivation were driven by wanting to generate profit, income, and assets. Farmers who scored highly on social motivation tended to be more strongly driven by family considerations and making a contribution to global food production.

The motivational profile is consistent with the literature (Maybery et al. 2005) and, importantly, supports the motivation factor structure of pastoralists proposed by Greiner et al. (2009b). That research had demonstrated a positive correlation between graziers' level of stewardship and lifestyle motivation and the adoption of best practice grazing land management, including the spelling of riparian areas and early de-stocking of the property when drought conditions were emerging. The recent choice experiment similarly found a significant positive influence of intrinsic interest in biodiversity — defined as the native animals and plants — and likelihood of participation in agri-environment schemes (Greiner 2015).

Table 13.1: What motivates pastoralists and graziers: Mean rating scores and factor loadings from a survey of northern Australian farmers (n=104).

Motivation items[1]	Mean rating score[2]	Motivation factors[3]		
		Stewardship and lifestyle motivation	Social motivation	Economic and financial motivation
Look after the natural assets of the property	4.5	0.8		
Pass on land in good condition	4.6	0.8		
Safeguard the property's natural assets	4.4	0.7		
Enjoy life and work on the property	4.6	0.7		
Improve resource/land condition	4.3	0.7		
Protect the environment	4.3	0.7		
Look after cattle	4.6	0.6		
Get satisfaction from living and working on the land	4.6	0.6		
Produce high-quality cattle	4.4	0.6		
Raise family on a grazing property	3.7		0.8	
Retire on the farm	2.5		0.6	
Ride horses/motorbikes/helicopters	3.1		0.6	
Put children through school/ university	4.0		0.6	
Step in ancestors' footsteps	2.3		0.6	
Produce beef to help feed the world population	3.9		0.5	

Motivation items[1]	Mean rating score[2]	Motivation factors[3]		
		Stewardship and lifestyle motivation	Social motivation	Economic and financial motivation
Earn a high income	3.1			0.7
Maximise company profit	4.0			0.7
Maximise cattle production from the land	4.0			0.7
Avoid years with very little or negative income	3.8			0.6
Build up land, wealth and assets	3.9			0.5
Be among the best in the industry	3.5			0.5
Run a profitable business	4.4			0.5
Eigenvalue		6.0	2.5	2.2
Cumulative (Eigenvalue)		6.0	8.6	10.7
% Total (variance)		27.4	11.5	9.9
Cumulative (%)		27.4	38.9	48.8

[1] Items in the table are sorted by factor association and loading value.

[2] Survey question: 'When you think about being a land owner/manager and pastoralist, how important are the following motivations to you?' A five-point response scale with 1='not at all important' to 5='extremely important'.

[3] Factor analysis conducted in Statistica 12, principal component extraction of factors, varimax orthogonal rotation, pairwise deletion of missing values, deletion of items with factor loadings <0.5.

Table 13.2: Attitudes of graziers and pastoralists towards biodiversity: Agreement with attitudinal statements and attitude factor scores from a survey of northern Australian farmers (n=104).

Attitudinal statements[1]	Mean score[2]	Biodiversity attitudes[3]		
		Stewardship ethic	Causes of/ solutions to biodiversity decline	Biodiversity on own property
As a landowner/land manager, I have an obligation to look after the native biodiversity and other natural assets on the property	4.5	0.8		
Caring for biodiversity is important to me personally	4.2	0.7		

Attitudinal statements[1]	Mean score[2]	Biodiversity attitudes[3]		
		Stewardship ethic	Causes of/ solutions to biodiversity decline	Biodiversity on own property
I take pleasure from seeing native biodiversity around	4.3	0.7		
Every pastoralist has a moral responsibility to look after the biodiversity and other natural assets on his/her land	4.3	0.6		
Abundance of certain native animals is an indicator of the health of the country	4.1	0.6		
Grazing plays a minor role in biodiversity decline compared to other pressures	3.5		0.8	
Statutory duty of care is sufficient to protect biodiversity	2.9		0.7	
Feral animals and plants pose a greater threat to native biodiversity than grazing	4.1		0.7	
Current national parks are sufficient to safeguard biodiversity of the savannas	2.9		0.6	
I have noticed a decline of native animals and plants on my property	1.9			0.7
Protecting endangered species on my property is easy	2.8			-0.7
It is relatively easy to safeguard native biodiversity on my property	3.4			-0.7
Eigenvalue		2.7	2.0	1.4
Cumulative (Eigenvalue)		2.7	4.8	6.2
% Total (variance)		19.5	14.5	10.3
Cumulative (%)		19.5	34.0	44.2

[1] Statements in the table are sorted by factor association and loading values.

[2] Survey question: 'How strongly do you agree or disagree with the following statements?' five-point response scale with 1='strongly disagree' to 5='strongly agree'.

[3] Factor analysis conducted in Statistica 12, principal component extraction of factors, varimax orthogonal rotation, pairwise deletion of missing values, deletion of items with factor loadings <0.5.

How farmers relate to biodiversity and what they think about agri-environment schemes influences likely participation in agri-environment schemes

The survey explored attitudes towards biodiversity and policy tools, including agri-environment schemes. Biodiversity attitudes were derived using factor analysis from the level of agreement with a Likert scale containing biodiversity related statements. The three-factor model provided a parsimonious construct of how farmers related to and thought about biodiversity (Table 13.2).

The factors captured the level of stewardship ethic, what farmers believed to be the causes of and solutions to biodiversity decline on pastoral land, and the extent to which they thought they could influence biodiversity on their properties. Farmers who scored highly on the 'stewardship ethic' factor attributed higher intrinsic value to biodiversity and believed landholders had a duty of care towards biodiversity. Farmers who scored highly on the 'causes of decline' factor tended to believe that biodiversity decline was mainly caused by factors other than grazing and that the formal conservation estate was sufficient to safeguard biodiversity. Farmers who scored highly on the 'biodiversity on own property' factor tended to notice a decline of biodiversity on their properties and did not think that safeguarding of biodiversity was a trivial task.

When these factors were included into the choice models, 'stewardship ethic' was shown to significantly and positively influence likely participation in agri-environment schemes (Greiner 2015). The level of 'stewardship ethic' as an attitude was significantly positively correlated with 'stewardship and lifestyle motivation' (p<0.001). Interpreted through the lens of Chapter 14, this can be interpreted as meaning farmers with higher stewardship ethic derive non-monetary private benefits from participation in agri-environment schemes, which reduces the financial incentive required. Accounting for these kinds of benefits when planning conservation can increase agri-environment schemes efficiency (for example, more biodiversity conservation for the same amount of program expenditure).

This research measured farmers' attitudes towards agri-environment schemes in terms of perceived effectiveness and found that, in general, farmers rated agri-environment schemes similarly favourably as income tax incentives and as more effective in incentivising more on-farm conservation than property planning, research, persuasion, and recognition measures (Table 13.3). Farmers who rated agri-environment schemes more favourably were found to be more likely to participate in contractual biodiversity conservation (Greiner 2015).

Table 13.3: Preferences for policy instruments and other measures: Perceived effectiveness based on a survey of northern Australian farmers (n=104).

Policy instruments and other items[1]	Mean score[2]	Standard deviation
Government investment in safeguarding/expanding overseas cattle markets	4.1	1.1
Income tax incentives	3.8	1.0
Financial incentives schemes (payments for ecosystem services) such as the ones explored in this research	3.8	1.1
Property management planning	3.5	1.1
Increased public acknowledgement of environmental achievements by graziers	3.4	1.2
More research into animals and grazing systems	3.4	1.1
Courses in grazing systems/grazing land management	3.3	1.2
More extension and consulting services offered on-farm	3.2	1.1
Debt-for-conservation swaps	3.1	1.3
Environmental management plans/systems	3.1	1.0
Industry organisations promoting the benefits of farm enterprise diversification	3.1	1.2
Voluntary (industry and regional) grazing code of practice	2.8	1.1
Increased peer recognition of grazier achievements (e.g. awards)	2.5	1.1
Community involvement (volunteers, schools) in on-ground works	2.4	1.2

[1] Items in the table sorted by mean effectiveness value.

[2] Survey question: 'How effective would the following measures be in helping you to undertake (more) conservation activities on your operation?'. A five-point response scale with 1='not at all effective' to 5='extremely effective'.

What this means for agri-environment scheme design

The empirical research summarised in this chapter suggests that agri-environment schemes are well liked by farmers and that participation in agri-environment schemes is readily considered once such programs are available to farmers. Participation decisions have been found to be subject to a number of factors. In addition to farm economic considerations about associated benefits and costs, a farmer's decision to participate in an agri-environment scheme at a certain incentive level is likely to be influenced by his or her like or dislike of contractual features, the trade-offs between contractual features, the reasons that drive him or her to be a farmer, and how he or she relates to the natural environment — in this case, biodiversity.

Figure 13.3: The brolga (*Grus rubicunda*) is one species that could benefit from environmental management in tropical savannas.
Source: Photo by Romy Greiner.

To ensure sufficient uptake to be effective and efficient, agri-environment schemes need to be carefully tailored to the contextual conditions. Of course, an agri-environment scheme is principally there to address an environmental problem and is only fit for purpose if the conservation requirements effectively address this problem. However, agri-environment schemes must equally succeed in engaging with the diversity of businesses, preferences, and attitudes of the target audience. The environmental success of agri-environment schemes depends on voluntary participation and farmers' willingness to participate, given the contractual conditions.

The challenge for the design and administration of agri-environment schemes is how to achieve effectiveness and efficiency of a program, and also consider matters of equity and procedural justice. On the basis of the research outlined herein, the following principles require consideration:

- Contracts need to stipulate conservation actions that meet the requirements of the biodiversity targeted for conservation. If some production can co-exist with conservation, farmers are significantly more willing to participate.

- Giving farmers a suite of contract options to choose from helps to entice landholder participation by responding to diverse cost structures (opportunity costs, risk premiums and transaction costs), diverse preferences for contract attributes, and diverse motivations and attitudes.

- Care needs to be taken to ensure that program features do not crowd out voluntary conservation actions that farmers may already be providing. This is particularly problematic in places such as northern Australia, where many farmers attribute a high value to biodiversity and are strongly motivated by stewardship considerations.

- In situations where intrinsic motivation is low and/or attitudes towards agri-environment schemes are unfavourable, complementary strategies are required to create an improved psychological foundation for agri-environment scheme participation. Complementary strategies can include information and extension efforts to articulate the values of biodiversity and demonstrate how farming impacts biodiversity, and showcase the effects of conservation actions. Regulatory and statutory requirements on landholders, for example in the form of an environmental duty of care (Greiner 2014b), are relevant supporting mechanisms.

- Allowing farmers to negotiate their land area contribution to an agri-environment scheme is important, particularly in an environment where individual operators are custodians of very large areas of land (the typical size of a pastoral station in northern Australia is around 2,500–10,000 km²). This gives rise to the scenario where a single farmer may control significant biodiversity assets on a portion of the property's area, so that engaging this single farmer in a conservation contract may be critical for safeguarding those biodiversity assets. The locality-specificity of biodiversity conservation combined with low number of potential services providers reduces opportunities for investors to implement agri-environment schemes in a competitive fashion (e.g. through environmental tenders). In such scenarios, negotiated approaches will be required with small numbers or single pastoralists who can achieve a desired conservation outcome. Successful negotiation will require a flexible approach to contract design.

- Biodiversity conservation on private land, funded by voluntary contractual arrangements, is an expensive way to do conservation, as the European experience shows (Batáry et al. 2015, p. 1014) and as the results of this research confirm. From the perspective of efficiency and permanency, inclusion of land into the formal conservation estate is preferable. However, agri-environment schemes can play a vital role of securing strategically important areas into a multi-tenure conservation system in the short- to medium-term.

Acknowledgements

This work was jointly funded by the National Environmental Research Program — Northern Australia Hub and Charles Darwin University. The author conducted the research at Charles Darwin University.

References

Ajzen, I. (1991) 'The theory of planned behaviour', *Organizational Behavior and Human Decision Processes* 50: 179–211.

Batáry, P., L.V. Dicks, D. Kleijn and W.J. Sutherland (2015) 'The role of agri-environment schemes in conservation and environmental management', *Conservation Biology*, 29: 1006–16.

Broch, S.W. and S.E. Vedel (2012) 'Using choice experiments to investigate the policy relevance of heterogeneity in farmer agri-environmental contract preferences', *Environmental and Resource Economics* 51: 561–81.

Burton, R.J.F. and G. Schwarz (2013) 'Result-oriented agri-environmental schemes in Europe and their potential for promoting behavioural change', *Land Use Policy* 30: 628–41.

Espinosa-Goded, M., J. Barreiro-Hurlé and E. Ruto (2010) 'What do farmers want from agri-environmental scheme design?: A choice experiment approach', *Journal of Agricultural Economics* 61: 259–73.

Garrod, G., E. Ruto, K. Willis and N. Powe (2012) 'Heterogeneity of preferences for the benefits of environmental stewardship: A latent-class approach', *Ecological Economics* 76: 104–11.

Greiner, R. (2014a) *Survey of north Australian graziers and pastoralists with choice experiment regarding participation in contractual biodiversity conservation*, dataset and questionnaire, Charles Darwin University. Available at: espace.cdu.edu.au/view/cdu:41684.

Greiner, R. (2014b) 'Environmental duty of care: From ethical principle towards a code of practice for the grazing industry in Queensland (Australia)', *Journal of Agricultural and Environmental Ethics* 27: 527–47.

Greiner, R. (2015) 'Factors influencing farmers' participation in contractual biodiversity conservation: A choice experiment with northern Australian pastoralists', *Australian Journal of Agricultural and Resource Economics* 58: 1–28.

Greiner, R., M.C.J. Bliemer and J. Ballweg (2014) 'Design considerations of a choice experiment to estimate likely participation by north Australian pastoralists in contractual on-farm biodiversity conservation', *Journal of Choice Modelling* 10: 34–45.

Greiner, R., I. Gordon and C. Cocklin (2009a) 'Ecosystem services from tropical savannas: Economic opportunities through payments for environmental services', *The Rangeland Journal* 31: 51–9.

Greiner, R. and D. Gregg (2011) 'Farmers' intrinsic motivations, barriers to the adoption of conservation practices and effectiveness of policy instruments: Empirical evidence from northern Australia', *Land Use Policy* 28: 257–65.

Greiner, R., L. Patterson and O. Miller (2009b) 'Motivations, risk perceptions and adoption of conservation practices by farmers', *Agricultural Systems* 99: 86–104.

Maybery, D., L. Crase and C. Gullifer (2005) 'Categorising farming values as economic, conservation and lifestyle', *Journal of Economic Psychology* 26: 59–72.

Pannell, D.J., G.R. Marshall, N. Barr, A. Curtis, F. Vanclay and R. Wilkinson (2006) 'Understanding and promoting adoption of conservation technologies by rural landholders', *Australian Journal of Experimental Agriculture* 46: 1407–24.

14

Accounting for private benefits in ecological restoration planning

Maksym Polyakov and David Pannell

Key lessons

- Selecting an effective ecological restoration project requires information about the levels of public and private net benefits that are likely to result from project implementation.
- Environmental assets on private land in agricultural landscapes may provide benefits that are valued by the landholders. The value of these benefits could be reflected in property sale prices.
- An extra hectare of native vegetation is valued more highly by the landholders of smaller properties and by the landholders of properties with smaller areas of native vegetation.
- Accounting for the private benefits generated by native vegetation when planning and targeting ecological restoration results in substantially greater biodiversity outcomes.

Introduction

Approximately 77 per cent of Australia's land area is managed by private landholders, which makes conservation on private lands an essential part of Australia's conservation strategy. Some examples of conservation initiatives on private land are the BushTender in Victoria, the Environmental Stewardship Project in New South Wales and Queensland, and the Whole of Paddock Rehabilitation (WOPR) scheme run by Greening Australia. Developing cost-effective ecological restoration programs on private lands is important, and it is crucial that the drivers of landholder participation in ecological restoration programs be identified (e.g. Blackmore and Doole 2013).

In this chapter, we explore several key lessons that would allow environmental managers to design effective ecological restoration programs by aligning landholders' private benefits with the public benefits of biodiversity conservation. We do this by identifying the circumstances in which native vegetation on private lands is likely to be valued by the landholder. This knowledge can be incorporated in the planning of ecological restoration programs and targeting of ecological restoration sites.

Public and private benefits

Environmental assets in rural landscapes provide a variety of benefits, or ecosystem services, to landholders (private benefits to the owners of the land in question) and to the public (public benefits to people other than the landholder). Private benefits of native vegetation include provision of shade for livestock, recreational opportunities, and increased amenity through improved aesthetics. Examples of public benefits provided by native vegetation include the provision of habitat for biodiversity, and regulation of water flows. The optimal allocation of rural land between intensive agricultural use, such as cropland or modified pasture, and native vegetation, which has conservation, amenity, and limited production value, depends on the balance between public and private benefits generated by the native vegetation, and the costs of land use change, such as ecological restoration, including the opportunity cost of foregone agricultural production.

Figure 14.1: A brown honeyeater feeding its young.
Source: Photo by Maksym Polyakov.

One of the ways to increase the provision of public benefits from agricultural landscapes is to use ecological restoration to reallocate land from intensive agriculture to conservation use. Designing effective policy instruments to implement natural resource management (NRM) actions, such as ecological restoration on private lands, requires information on private benefits generated by these management actions. Pannell (2008) developed a framework to guide the choice between policy mechanisms, based on the levels of public and private net benefits likely to result from proposed management actions (see Chapter 18). The framework highlights the importance of targeting funds in environmental programs to selected spatial targeting areas, based on the likely levels of public and private net benefits. For example, selecting ecological restoration projects that provide public benefits and modest private benefits would provide high value for money, because policy mechanisms are likely to influence behaviour at relatively low public cost (Polyakov et al. 2015).

Environmental benefits valued by the landholders

Environmental assets located on private properties, such as native vegetation, may generate private benefits that could be reflected in property values. These values can be estimated using the hedonic pricing method. While there have been many studies that value environmental assets on public lands, few studies have investigated the values of ecosystem services generated by environmental assets located on private rural lands. For example, Ma and Swinton (2011) found that environmental assets, such as forest, wetlands, and streams, located both on the property and in the surrounding landscape, increase the value of rural properties in Michigan. Walpole and Lockwood (1999) measured the effect of native vegetation on the values of rural properties in north east Victoria and southern New South Wales. While they did not find a measureable impact on property value when native vegetation covered less than 50 per cent of property area, coverage above 50 per cent decreased property value. Polyakov et al. (2013) estimated that the native vegetation on rural lifestyle properties in Victoria has a positive and diminishing marginal value, with the property value maximised when the proportion of area occupied by native vegetation is approximately 40 per cent, which increases property value by 10.5 per cent relative to the value of a similar property without native vegetation. However, when the area occupied by native vegetation exceeds 80 per cent of the property, the value of the property is reduced to less than the value associated with no native vegetation. Since the current extent of native vegetation is lower than the extent that would maximise its amenity value to many landholders (Polyakov et al. 2013), the welfare value to people living in the landscape may be improved by restoring native vegetation on cleared lands.

These examples demonstrate that environmental assets in agricultural landscapes, which generate supporting, regulating, and cultural ecosystem services, provide private benefits to the landholders. These benefits are reflected in increased property values. Ecological restoration on private land can increase both the welfare of landholders and the provision of public benefits.

Property size and extent of native vegetation matters

Native vegetation on private land provides a range of cultural, recreational, and aesthetic amenity benefits to landholders. On rural properties, the retention of native vegetation is generally in conflict with agricultural practices used to produce food and fibre. As with any limited resource, both native vegetation and agricultural land are likely to exhibit diminishing marginal values. This means that the value of the last hectare of land is lower than the value of the second last hectare. There could be an optimal combination of native vegetation and agricultural land on private rural property that maximises the total benefit a landholder derives from the property, which would be reflected in a higher property value. The optimal combination of these land uses depends on the landholder's goals and preferences (see Chapter 13). It is likely that the value of native vegetation varies across the spectrum of landholders, such as full-time farmers, part-time farmers, and lifestyle farmers. For example, owners of lifestyle properties have relatively strong preferences for amenity values of native vegetation, while owners of large agricultural properties have preference for the production value of agricultural land.

Polyakov et al. (2015) used data from around 7,500 rural properties in north central Victoria sold between 1990 and 2011 to estimate the value that a variety of rural landholders place on native vegetation on their properties. They used property size as a proxy for landholder type to model differences in the values of native vegetation across the range of property types in a multifunctional rural landscape. They found that the value of native vegetation is smaller on larger properties, which are associated with production-oriented farmers, and larger on smaller properties, which are associated with lifestyle landholders. The value of additional native vegetation is higher on the properties which currently have little native vegetation, and decreases with greater areas of native vegetation. This implies that increasing the area of native vegetation on a property increases its value up to an optimal point (see Figure 14.2), after which additional native vegetation decreases the property value. The proportion of native vegetation that maximises the property value, as well as the extent to which it increases the property value, varies across property sizes (see Figure 14.2). This reflects landholders' targets and priorities. For example, a 1 hectare property would reach its highest value when native vegetation covers about 45 per cent of its area. Such a property would be worth approximately 25 per cent more than a similar property without any native vegetation. Further increase of the area of native vegetation would decrease the property value relative to the optimum, and when the proportion of native vegetation exceeds 90 per cent of the property area, its value becomes lower than the value of similar property without any native vegetation. As the property size increases, the optimal proportion of native vegetation becomes lower and native vegetation has a smaller impact on property values. The optimal proportions of native vegetation for 10 hectare, 100 hectare, and 1,000 hectare properties are estimated to be 37 per cent, 29 per cent, and 20 per cent, respectively. These proportions would increase property values by, respectively, 16 per cent, 9 per cent, and 5 per cent relative to the values of similar properties without any native vegetation. Adding more native vegetation after reaching these optimal proportions would decrease property values: after proportions of native vegetation reach 64 per cent, 58 per cent, and 40 per cent, respectively, values of these properties would become lower than the values of similar, fully cleared properties.

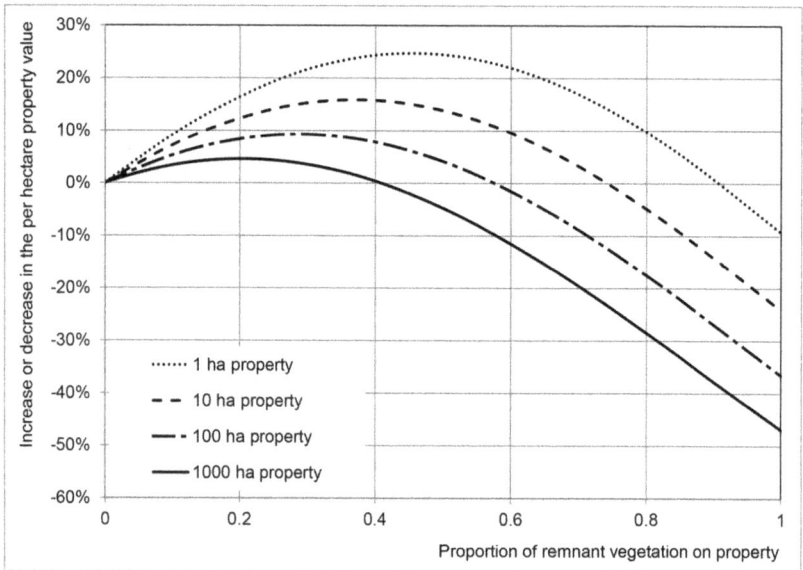

Figure 14.2: Effect of the proportion of native vegetation on land value by property size over time.

Source: Maksym Polyakov, David J. Pannell, Ram Pandit, Sorada Tapsuwan, Geoff Park, 'Capitalized Amenity Value of Native Vegetation in a Multifunctional Rural Landscape', *American Journal of Agricultural Economics*, 2014, 97(1): 299–314, by permission of Oxford University Press.

It is important to account for private benefits when planning ecological restoration

A large proportion of the costs of ecological restoration projects consist of opportunity costs, which in this context is the loss of income that occurs when land use is changed from agriculture or other productive use to conservation. Traditionally, opportunity costs of ecological restoration in agricultural landscapes have been accounted for using land values (Westphal et al. 2007) or the capitalised revenue of agricultural production (Crossman and Bryan 2006). The latter is aggregated present value of future net income from the land, which should approximately match the land's sale value if the only values that matter are commercial values. Using these values to represent the opportunity cost of restoration implies that when a parcel of land is acquired for an ecological restoration project, the landholder loses all private benefits generated by this parcel. However, when ecological restoration is conducted on private land with the landholder retaining ownership of the land, the landholder captures the benefits generated by the restored ecosystem, such as amenity from native vegetation. This reduces the opportunity cost and overall cost of the restoration project. Furthermore, because the value of privately captured benefits of ecosystem services varies with property type, size, and the amount of existing native vegetation on the property (see Figure 14.2), the opportunity costs of ecological restoration would be lower on smaller properties and properties with little native vegetation. Ignoring private benefits in calculation of opportunity costs of ecological restoration on private lands could lead to misleading recommendations for ecological restoration planning.

In a study of Mt Alexander Shire in Victoria, Polyakov and Pannell (2014) tested whether ignoring private benefits in calculating the opportunity cost resulted in different optimal (from the point of view of maximising public benefits) spatial allocation of ecological restoration and different biodiversity outcomes. They compared optimal ecological restoration strategies resulting from two different assumptions about opportunity cost. In one scenario, opportunity cost was based on the land value, while in another scenario, opportunity cost was estimated by taking into account the land value and private benefits of native vegetation. The scenario accounting for private benefits of ecological restoration gave a substantially (up to 75 per cent) better biodiversity outcome than the scenario using opportunity cost based on only land values. The spatial pattern of ecological restoration in these scenarios differed substantially. In the scenario that uses land values for opportunity cost, ecological restoration is selected on larger properties, which have lower per hectare land values. In the scenario that takes into account private benefits of native vegetation, ecological restoration takes place on smaller properties (lifestyle landholders). The land value of these properties is higher, but converting part of the property to native vegetation does not decrease property value. As a result, allocation of ecological restoration shifts towards smaller properties, even though transaction costs and overhead costs of implementing smaller ecological restoration projects are higher. This outcome is consistent with the findings of Race et al. (2010) that lifestyle landholders and part-time farmers undertake a considerable amount of work to revegetate and enhance native vegetation.

These results show that in order to avoid providing misleading recommendations to environmental managers, it is important to take into account amenity values of native vegetation and variable opportunity cost when prioritising ecological restoration. This is especially important in multifunctional landscapes with heterogeneous landholders.

Conclusion

This chapter summarises lessons from recent studies of the private benefits of environmental assets in rural landscapes, in the context of ecological restoration. Ecosystem services of the environmental assets, such as native vegetation on private lands, generate private benefits that are captured by the landholders. The value of private benefits generated by an extra hectare of native vegetation depends on the property type and area of native vegetation currently on the property. This information on private benefits from restoration is relevant to the decisions about the targeting of ecological restoration because private benefits of ecological restoration can reduce the public cost. Specifically, landholders with high marginal private benefits of revegetation would be more willing to participate in revegetation programs. Targeting those landholders would provide better value for money because the program could be implemented at lower cost and with higher probability of success. Accounting for private benefits in planning and targeting restoration would result in restoration programs delivering greater benefits.

Acknowledgements

This research was conducted with the support of funding from the Australian Research Council Centre of Excellence for Environmental Decisions and the Australian National Environmental Research Program's Environmental Decisions Hub.

References

Blackmore, L. and G.J. Doole (2013) 'Drivers of landholder participation in tender programs for Australian biodiversity conservation', *Environmental Science and Policy* 33: 143–53.

Crossman, N.D. and B.A. Bryan (2006) 'Systematic landscape restoration using integer programming', *Biological Conservation,* 128: 369–83.

Ma, S. and S.M. Swinton (2011) 'Valuation of ecosystem services from rural landscapes using agricultural land prices', *Ecological Economics* 70: 1649–59.

Pannell, D.J. (2008) 'Public benefits, private benefits, and policy mechanism choice for land-use change for environmental benefits', *Land Economics* 84: 225–40.

Polyakov, M. and D.J. Pannell (2014) 'Accounting for private benefits in ecological restoration planning', presentation to Agricultural and Applied Economics Association 2014 Annual Meeting, July 27–29, Minneapolis, Minnesota.

Polyakov, M., D.J. Pannell, R. Pandit, S. Tapsuwan and G. Park (2013) 'Valuing environmental assets on rural lifestyle properties', *Agricultural and Resource Economics Review* 42: 159–75.

Polyakov, M., D.J. Pannell, R. Pandit, S. Tapsuwan and G. Park (2015) 'Capitalized amenity value of native vegetation in a multifunctional rural landscape', *American Journal of Agricultural Economics* 97: 299–314.

Race, D., R. Sample, A. Curtis and S. McDonald (2010) *Management of native vegetation on private land: Perspectives of landholders and NRM program managers in northern Victoria*, a report to Landscape Logic, Commonwealth Environmental Research Program, Canberra.

Walpole, S.C. and M. Lockwood (1999) 'Influence of remnant native vegetation on rural land values: A hedonic pricing application', presentation to the Australian and New Zealand Agricultural and Resource Societies 1999 Joint Conference, Christchurch, New Zealand.

Westphal, M.I., S.A. Field and H.P. Possingham (2007) 'Optimizing landscape configuration: A case study of woodland birds in the Mount Lofty Ranges, South Australia', *Landscape and Urban Planning* 81: 56–66.

Part III.
Planning, doing and learning

15

Defining and designing cost-effective agri-environment schemes

Dean Ansell

Key lessons

- Agri-environment schemes are often highly variable in both their economic cost and biodiversity benefit, creating the potential for significant inefficiencies in conservation expenditure.
- Evaluation of the cost-effectiveness of agri-environment schemes can identify opportunities to significantly improve the conservation gains with the available resources, however, such evaluations are uncommon.
- Simple economic evaluation tools can be applied by researchers or policymakers, using minimal economic data, to compare the cost-effectiveness of agri-environment schemes at different scales and at stages through the implementation process.

Introduction

Over the past decade, concerns have been raised regarding the effectiveness of agri-environment schemes in conserving biodiversity. Studies have shown that the success of these schemes is highly variable, ranging from strong positive biodiversity benefits to neutral and even negative consequences. With global biodiversity declining dramatically and further threatened by agricultural intensification, a focus on the most effective strategies for conservation is critical. This issue is even more pertinent given that the funds available for biodiversity conservation are not sufficient to address the scale of the problem, and so agri-environment schemes are in competition with other conservation activities for limited resources. It is critical, therefore, that the cost-effectiveness of agri-environment schemes are maximised to increase the biodiversity benefits obtained with available resources.

Figure 15.1: The cost-effectiveness of agri-environment schemes can be influenced by many factors, from the location of sites to the specific conservation techniques used.

Source: Photo by Brisbane City Council available at www.flickr.com/photos/brisbanecity council/7926277216 under a Creative Commons Attribution 2.0.

Typically, evaluation of agri-environment schemes has been dominated by ecological or economic perspectives (Uthes and Matzdorf 2013). There has been comparatively little attention given to the cost-effectiveness of these schemes. A recent review of 239 studies on the effectiveness of agri-environment schemes around the world found that fewer than 15 per cent considered economic costs in the evaluation

(Ansell et al. in preparation). This is surprising, given the scale of the public expenditure in agri-environment schemes and increasing recognition of the biodiversity benefits that can be achieved through consideration of economic costs in the conservation planning process (see Chapter 17 by Fiona Gibson and colleagues).

This chapter explores issues of the effectiveness and efficiency of agri-environment schemes, first defining cost-effectiveness in the context of such schemes, and providing an overview of common evaluation approaches. I conclude with a discussion on the outcomes of previous evaluations of the cost-effectiveness of agri-environment schemes, highlighting key aspects relevant to the design and implementation of agri-environment schemes in Australia.

What is cost-effectiveness?

Cost-effectiveness refers to the relative efficiency of an action in achieving an outcome. It can be expressed as the total cost of producing a single unit of benefit (i.e. cost/benefit), or alternatively as the total benefit produced per unit of cost (i.e. benefit/cost) (Wätzold and Schwerdtner 2005). Both approaches generate a ratio, referred to as the cost-effectiveness ratio or benefit–cost ratio, which forms the basis of a cost-effectiveness analysis. The ratio allows one to compare the efficiency of alternative actions. Note that while I focus here on evaluation approaches involving non-monetary measures of conservation benefit, as opposed to methods that assign a monetary value to the effectiveness measure, the concepts discussed apply generally across both approaches.

When applied to the evaluation of biodiversity benefits of agri-environment schemes, we can take cost-effectiveness as the biodiversity benefit produced per unit of cost (or, alternatively, cost per biodiversity unit). Comparison of the cost-effectiveness of different agri-environment schemes allows identification of those that provide the greatest biodiversity benefit per dollar spent.

Variation in both the economic costs and effectiveness of conservation activities give rise to differences in the cost-effectiveness of agri-environment schemes (Wätzold and Schwerdtner 2005). Agricultural ventures are rarely static in time and space, with farming practices, production intensity, and commodity choice varying according to various external market factors (Barraquand

and Martinet 2011). This gives rise to significant variation in the opportunity costs of conservation on farmland at multiple spatial and temporal scales. Similarly, the effectiveness of agri-environment schemes in conserving biodiversity is highly variable, both temporally and spatially, and is influenced by factors at the field or farm scale (e.g. site size, management history) as well as at the landscape or regional scales (e.g. surrounding land use, connectivity, climate) (e.g. Concepción and Díaz 2011). Effectiveness also varies between taxa, with some schemes providing positive conservation outcomes for some taxa while providing no benefit or even negative outcomes for other taxa (Kleijn et al. 2006).

Complex interdependencies also exist between the effectiveness and cost of conservation activities in agri-environment schemes. For example, the overall cost of an agri-environment scheme is strongly influenced by the configuration (i.e. size and shape) of the particular field, with larger sites incurring a higher opportunity cost to the landholder, in turn requiring an increased payment rate, and often requiring increased materials. This can also influence biodiversity outcomes, with factors such as field size and shape shown to be important determinants of conservation effectiveness (Conover et al. 2011). This variation in costs and effectiveness, and the complex interactions between the two, create the potential for significant inefficiencies in conservation expenditure. Simple evaluation of the cost-effectiveness of agri-environment schemes can reveal factors driving conservation efficiency and identify opportunities to maximise the conservation benefits from investments.

Cost-effectiveness in practice

A critical step in the evaluation of the cost-effectiveness of agri-environment expenditure is the assessment of the costs of the scheme itself. The costs associated with agri-environment schemes can be categorised as acquisition (e.g. land rent), management (e.g. site establishment, maintenance), transaction (e.g. negotiation, legal), and opportunity costs (e.g. forgone agricultural production) (Naidoo and Ricketts 2006). Consideration of the latter is particularly critical, as it can influence the design, uptake, and ultimately the effectiveness of biodiversity conservation in farmland, but is often omitted from project evaluations.

Evaluations can use realised or actual costs (Klimek et al. 2008), or estimated costs based on market rates, averages, or surrogates (Bamière et al. 2013). While evaluation based on actual costs provides improved accuracy, such information is not always readily available. Naidoo and Ricketts (2006) provide an overview of approaches for the estimation of common cost components of biodiversity conservation. Evaluations should attempt to take account of the full costs (and benefits) of agri-environment schemes (Bamière et al. 2013).

The other key component in the evaluation of cost-effectiveness is obviously the measure of the benefit or the effectiveness of conservation. As the numerator in most cost-effectiveness equations, it can strongly influence the outcomes of the evaluation and therefore careful selection is critical. Evaluation can use direct or field-based measures of effectiveness such as changes in single species abundance or density, or, alternatively, look at measures of community diversity (e.g. Ulber et al. 2011). Measures of habitat area or quality have also been used, either as direct measures of agri-environment schemes effectiveness (e.g. Wynn 2002), or as surrogate measures of broader biodiversity benefits (e.g. Hansen 2007). Thompson et al. (1999) use area of land enrolled in the particular agri-environment schemes under review as a proxy for effectiveness.

Many evaluations, particularly those carried out at the planning stages of agri-environment schemes (i.e. ex ante, see below) are based on modelled or predicted outcomes as measures of effectiveness. For example, Bamière et al. (2013) use spatial modelling to assess the efficiency of agri-environment policies by focusing on the spatial configuration of farm land for habitat conservation, specifically aiming for a random mosaic of sites, noting that such a configuration is more effective in the conservation of certain species, such as their model species, the little bustard, which depends on a mosaic of agricultural land use (i.e. crops, grassland).

Except where the particular objectives of the agri-environment schemes or research question dictates the use of a specific measure of effectiveness — for example, changes in the abundance of a species or in the area of a certain habitat — the researcher will be faced with the difficult task of selecting a suitable measure to capture, to the extent possible, the extent of the biodiversity benefits resulting from the scheme. In such cases, multiple benefits can be captured in a metric that can then be used in economic analysis. Multi-criteria analysis approaches have also been used to combine multiple disparate environmental values into single measures (e.g. Hajkowicz and Collins 2009). (In Chapter 17, Fiona Gibson and David Pannell discuss the consequences of using the wrong metric, while in Chapter 20 Phil Gibbons provides an overview of the development of metrics.)

Irrespective of the particular effectiveness measure selected, careful consideration should be given to the means of collecting that information and its expression within the cost-effectiveness evaluation. Experimental design is critical in the assessment of cost-effectiveness. Failure to account for conservation status in the absence of the treatment (i.e. control or counterfactual) can lead to inflated measures of benefit and cost-effectiveness (a topic discussed by Duncan and Reich in Chapter 19). Kleijn and Sutherland (2003) found significant shortcomings in the design of studies evaluating the effectiveness of European agri-environment schemes, stemming largely from either poor, or absent, controls. The authors propose a number of remedies, including the use of baseline data, comparison of trends in treatments and controls, and use of carefully selected treatment and control site pairs. Of particular importance is the use of conservation gain (i.e. the difference in the biodiversity value between the treatment and control) as the measure of conservation benefit, rather than absolute values. This provides a more accurate measure of the benefit that has been purchased with the investment and controls for differences in the baseline condition or value (Maron et al. 2013).

Cost-effectiveness can be considered at a variety of scales in the agri-environment schemes process. For example, we can consider the efficiency of different agri-environment policies in achieving environmental outcomes at a broad scale. Bamière et al. (2013) use a modelling approach to compare the cost-efficiency of three different agri-environment schemes, each using a different incentive mechanism, in achieving a specific objective for the conservation of little bustard habitat in French farmland. In contrast, we can compare the cost-effectiveness of specific measures in achieving their biodiversity objectives. Wilson et al. (2007) evaluate the cost-effectiveness of two different conservation activities (a low cost habitat preservation option and a high cost habitat restoration option) aimed at improving wading bird populations in southern England under the Environmentally Sensitive Areas scheme. They find that, despite the habitat restoration measure costing 50 per cent more per hectare than the habitat preservation measure, the return on investment from the higher cost option, measured as cost per breeding pair of waders, was more than 90 per cent higher than the low-cost option. This provides a strong example of the power of a simple cost-effectiveness analysis in comparing the efficiency of different conservation activities. The choice of scale for evaluation should be appropriate for the research or policy question, and will influence the detail or resolution of the ecological and economic information required in the evaluation.

Box 15.1: Before, during and after — timing of AES evaluation.

We can also consider the cost-effectiveness of agri-environment schemes at different time stages throughout the process. Evaluations carried out prior to the implementation are referred to as ex ante evaluations and can provide important input into scheme design and implementation. Such evaluations provide the opportunity to optimise the efficiency and biodiversity benefits of agri-environment schemes investments. Van der Horst (2007) assessed the efficiency gains from spatial targeting of a woodland agri-environment schemes and found biodiversity gains of 1.6–2.1 times greater than that achieved through the untargeted scheme. White and Sadler (2011) achieve a 17 per cent budget saving through the use of conservation contracts based on variable payments tailored to outcomes achieved on individual enrolled farms compared to traditional fixed-price contracts.

Evaluations can also be carried out during (in media res) or upon completion (ex post) of a scheme. In contrast to ex ante evaluations, which typically involve modelling of predicted biodiversity benefits and costs, such evaluations can use realised benefits and actual costs as inputs, provide a retrospective assessment of the efficiency of expenditure, and identify improvements for future programs. Both ex ante and ex post evaluations provide useful information about the biodiversity benefits of agri-environment schemes. While the use of predicted biodiversity benefit and cost information in ex ante evaluations may provide less accurate information than approaches using realised benefits and actual costs (i.e. ex ante evaluations) (Boardman et al. 2010), conducting evaluations at this stage may improve the efficiency of an agri-environment scheme before funding is expensed. In contrast, ex post evaluations, while providing information to improve the efficiency of future expenditure, can be hampered by limited availability of financial data and methodological issues around the measurement of biodiversity gains.

Despite these shortcomings, both approaches can contribute to the refinement of agri-environment schemes and increase the biodiversity gains and efficiency of agri-environment expenditure. Ultimately, the choice of evaluation approach may be determined by financial and logistical constraints.

Lessons learnt

While we may be tempted to think that the more we spend on agri-environment schemes, the better the biodiversity outcomes, evaluations reveal the relationship between the two is anything but straightforward. While some studies support this concept by demonstrating higher levels of conservation benefit with increasing expenditure (Barraquand and Martinet 2011; Wilson et al. 2007), others reveal more complex relationships. For example, in an ex post evaluation of the Scottish Woodland Grants Scheme, which aimed to improve priority habitats in farmland, Wynn (2002) found wide variation in cost, biodiversity benefit, and cost-effectiveness across different farm types.

Shining a light on the economics of biodiversity conservation in farming landscapes can reveal some ugly truths that would otherwise not be uncovered by traditional ecological evaluations. Examples include the prevalence of significant windfall effects in agri-environment schemes, where farmers receive payments for environmental services or biodiversity outcomes that would have occurred regardless of whether the scheme was implemented (Chabé-Ferret and Subervie 2013; Sierra and Russman 2006; Ulber et al. 2011), and a reliance on agri-environment schemes payments for farm income (Pietzsch et al. 2013). Recent modelling of the cost-effectiveness of habitat restoration on Australian farmland suggests that our current focus on restoring remnant habitats, as is the focus of the Australian Government's largest agri-environmental scheme, the Environmental Stewardship Scheme (see Chapter 3 by Burns and colleagues), is suboptimal, with revegetation of cleared areas demonstrating higher biodiversity gains per dollar spent (Jellinek et al. 2014).

While it is important that seemingly negative research outcomes such as these be evaluated and communicated, there is a potential risk of perverse conservation outcomes where seemingly adverse economic results drive policy decisions (i.e. cancellation of programs) at the expense of important biodiversity values or priorities. The challenge is in maintaining perspective in evaluating the cost-effectiveness of agri-environment schemes and assessing the outcomes of such evaluations in the context of the scheme's overall biodiversity objectives.

Evaluation can also provide important lessons for the design of future conservation programs. Many agri-environment schemes use a simple incentive system, where payments to farmers are based on fixed rates per hectare. Such approaches are relatively easy to administer, but risk significant inefficiency though overcompensation of farmers otherwise willing to accept a lower price for conservation (Klimek et al. 2008). This can also exacerbate the problem of marginal, low productivity areas dominating the enrolled land as farmers seek to minimise opportunity cost and maximise returns from enrolment (Bamière et al. 2013).

Several studies demonstrate the efficiency gains that can be achieved through more complex delivery mechanisms, such as auction-based and payment-by-results type systems (e.g. Barraquand and Martinet 2011; Thompson et al. 1999; Klimek et al. 2008). Stoneham et al. (2003) compared the outcomes of a pilot auction for the Victorian BushTender scheme and found such an approach would achieve the same biodiversity outcomes at a cost seven times less than those achieved using a fixed-rate incentive payment. The increased efficiency of these approaches, however, must be balanced against the higher administrative or transaction costs associated with their implementation (Klimek et al. 2008; White and Sadler 2011).

Conclusion

It is unfortunate that better use is not made of simple tools of economic evaluation in the planning and assessment of conservation expenditure in agricultural land. By focusing only on biological or ecological aspects in our evaluations, we miss opportunities to significantly increase the biodiversity benefits that can be achieved with the limited funding available. As demands to feed a growing population place even greater pressure on biodiversity in agricultural landscapes and the conservation purse strings tighten, maximising the efficiency of our conservation dollar becomes even more critical. Understanding the cost-effectiveness of our agri-environment investments is a critical step towards meeting this aim.

References

Ansell, D., D. Freudenberger, N. Munro and P. Gibbons (in preparation) *The cost-effectiveness of agri-environment schemes: a systematic review.*

Bamière, L., M. David and V. Vermont (2013) 'Agri-environmental policies for biodiversity when the spatial pattern of the reserve matters', *Ecological Economics* 85: 97–104.

Barraquand, F. and V. Martinet (2011) 'Biological conservation in dynamic agricultural landscapes: Effectiveness of public policies and trade-offs with agricultural production', *Ecological Economics* 70(5): 910–20.

Boardman, A., et al. (2010) *Cost-benefit analysis*, fourth edition, Prentice Hall, Upper Saddle River, NJ.

Chabé-Ferret, S. and J. Subervie (2013) 'How much green for the buck?: Estimating additional and windfall effects of French agro-environmental schemes by DID-matching', *Journal of Environmental Economics and Management* 65(1): 12–27.

Concepción, E.D. and M. Díaz (2011) 'Field, landscape and regional effects of farmland management on specialist open-land birds: Does body size matter?' *Agriculture, Ecosystems and Environment* 142: 303–10.

Conover, R.R., S.J. Dinsmore and L.W. Burger (2011) 'Effects of conservation practices on bird nest density and survival in intensive agriculture', *Agriculture, Ecosystems and Environment* 141(1–2): 126–32.

Hajkowicz, S. and K. Collins (2009) 'Measuring the benefits of environmental stewardship in rural landscapes', *Landscape and Urban Planning* 93(2): 93–102.

Hansen, L. (2007) 'Conservation reserve program: Environmental benefits update', *Agricultural and Resource Economics Review* 2: 267–80.

Herzog, F. et al. (2005) 'Effect of ecological compensation areas on floristic and breeding bird diversity in Swiss agricultural landscapes', *Agriculture, Ecosystems and Environment* 108(3): 189–204.

Hinsley, S.H.A. et al. (2010) 'Testing agri-environment delivery for farmland birds at the farm scale: The Hillesden experiment', *Ibis* 152: 500–14.

Jellinek, S., L. Rumpff, D.A. Driscoll, K.M. Parris and B.A. Wintle (2014) 'Modelling the benefits of habitat restoration in socio-ecological systems', *Biological Conservation* 169: 60–7.

Kleijn, D., R. Baquero and Y. Clough (2006) 'Mixed biodiversity benefits of agri-environment schemes in five European countries', *Ecology Letters*, 9(3): 243–54.

Kleijn, D. and W.J. Sutherland (2003) 'How effective are European agri-environment schemes in conserving and promoting biodiversity?', *Journal of Applied Ecology* 40(6): 947–969.

Klimek, S. et al. (2008) 'Rewarding farmers for delivering vascular plant diversity in managed grasslands: A transdisciplinary case-study approach', *Biological Conservation* 141(11): 2888–97.

Knop, E. and D. Kleijn (2006) 'Effectiveness of the Swiss agri-environment scheme in promoting biodiversity', *Journal of Applied Ecology* 43: 120–7.

Konvicka, M. et al. (2008) 'How too much care kills species: Grassland reserves, agri-environmental schemes and extinction of Colias myrmidone (Lepidoptera : Pieridae) from its former stronghold', *Journal of Insect Conservation* 12(5): 519–25.

Maron, M., J. Rhodes and P. Gibbons (2013) 'Calculating the benefit of conservation actions', *Conservation Letters* 6(5): 359–67.

McCarthy, D., et al. (2012) 'Financial costs of meeting global biodiversity conservation targets: current spending and unmet needs', *Science* 338: 946–9.

Naidoo, R. and T.H. Ricketts (2006) 'Mapping the economic costs and benefits of conservation' *PLoS Biology* 4(11): e360.

Pietzsch, D., et al. (2013) 'Low-intensity husbandry as a cost-efficient way to preserve dry grasslands', *Landscape Research* 38(4): 523–39.

Sierra, R. and E. Russman (2006) 'On the efficiency of environmental service payments: A forest conservation assessment in the Osa Peninsula, Costa Rica', *Ecological Economics* 59: 131–41.

Stoneham, G., V. Chaudhri, A. Ha and L. Strappazzon (2003) 'Auctions for conservation contracts: An empirical examination of Victoria's BushTender trial', *Australian Journal of Agricultural and Resource Economics* 47: 477–500. DOI:10.1111/j.1467-8489.2003.

Thompson, S., A. Larcom and J.T. Lee (1999) 'Restoring and enhancing rare and threatened habitats under agri-environment agreements: A case study of the Chiltern Hills area of outstanding natural beauty, UK', *Land Use Policy* 16(2): 93–105.

Ulber, L., et al. (2011) 'Implementing and evaluating the effectiveness of a payment scheme for environmental services from agricultural land', *Environmental Conservation* 38(4): 464–72.

Uthes, S. and B. Matzdorf (2013) 'Studies on agri-environmental measures: A survey of the literature', *Environmental management* 51(1): 251–66.

Van der Horst, D. (2007) 'Assessing the efficiency gains of improved spatial targeting of policy interventions: The example of an agri-environmental scheme', *Journal of Environmental Management* 85: 1076–87.

Wätzold, F. and K. Schwerdtner (2005) 'Why be wasteful when preserving a valuable resource?: A review article on the cost-effectiveness of European biodiversity conservation policy', *Biological Conservation* 123: 327–38.

White, B. and R. Sadler (2011) 'Optimal conservation investment for a biodiversity-rich agricultural landscape', *Australian Journal of Agricultural and Resource Economics* 56: 1–21.

Whittingham, M.J. (2007) 'Will agri-environment schemes deliver substantial biodiversity gain, and if not why not?', *Journal of Applied Ecology* 44: 1–5.

Wilson, A., J. Vickery and C. Pendlebury (2007) 'Agri-environment schemes as a tool for reversing declining populations of grassland waders: Mixed benefits from environmentally sensitive areas in England', *Biological Conservation* 136(1): 128–35.

Wynn, G. (2002) 'The cost-effectiveness of biodiversity management: A comparison of farm types in extensively farmed areas of Scotland', *Journal of Environmental Planning and Management* 45(6): 37–41.

16

Transaction costs in agri-environment schemes

Stuart Whitten and Anthea Coggan

Key lessons

- Transaction costs of agri-environment schemes include the time, effort and expense of gathering information, identifying projects, negotiating contracts, and monitoring and compliance.
- They are incurred by participants, scheme proponents and administrators and can be significant, impacting not only on total scheme costs, but also on efficiency.
- Transaction costs are directly related to both scheme design and scheme implementation.
- Considering transaction costs does not necessarily mean reducing them — indeed, efficient program design may require increased transaction costs in order to more confidently deliver the desired outcome.

Transaction costs — a necessary evil?

Agri-environmental schemes, which are designed to support private land managers in delivering positive environmental outcomes, involve a range of costs to government and landholders. These include the direct costs from implementing desired agri-environmental management actions (materials, labour, and equipment), opportunity costs from changes to agricultural production, transformation costs (physical or other changes to the business to allow the management actions to be implemented), and transaction costs (arising from program design, implementation, and management). While direct costs are generally highly visible and well reported, transformation and transaction costs are often less obvious and are neglected in analyses. Transaction costs are a particularly pervasive, yet relatively recent concept in economics, important in the design and implementation of agri-environment schemes.

Figure 16.1: There are numerous transaction costs associated with agri-environment schemes, including compliance monitoring and program evaluation.

Source: Photo by Declan Feeney available at www.flickr.com/photos/ardboline/9797787304 under a Creative Commons Attribution 2.0.

So what are these transaction costs? There are three overlapping definitions that apply somewhat differently depending on the particular environmental issue that a scheme is concerned with. From the perspective of the organisations involved in an agri-environment scheme, transaction costs relate to the costs of gathering information, negotiating or otherwise identifying who to pay, who to be paid by, and how much, as well as contracting, monitoring, and enforcement

(Williamson 1998). These costs apply to the scheme proponent (usually government agencies or third parties engaged by government) and to firms, individuals (i.e. landholders), or other agents engaging in the scheme (McCann et al. 2005). They are generally held to include the direct costs of designing the scheme, but a narrower definition applied by North (1990, pp. 1–35) limits transaction costs to the costs of specifying and supporting the contract. This definition is most relevant in direct comparison of existing agri-environment schemes designed for the same purpose, where the costs of designing the instrument can be considered sunk or legacy costs, and evaluation only applies to future efficiency.

The third definition applies when social coordination is required to produce the desired outcome — such as biodiversity corridors or management of a common water resource (Reeson et al. 2011). The costs of social coordination, through collective action, consultative planning or other measures would then be included as a transaction cost (Ostrom 1990). While such coordinated action is by no means unusual in agri-environmental settings, most schemes are designed and intended to interact with individuals separately, even where a socially coordinated outcome is intended. These schemes are also sufficiently novel that they are not generally available off the shelf and involve at least some investment in refining their design and implementation for a particular context. The relevant transaction costs in the most common agri-environmental settings include design, implementation, and administration costs across proponents and participants. Design and implementation costs do not generally involve changes to the institutional environment and legal settings when the scheme is limited to payments to landholders for existing rights, but may if the scheme involves creation of tradeable rights that are then purchased — such as in new water markets or biodiversity offset schemes.

Transaction costs may be monetary costs (e.g. administrative staff, legal advice on contracts, cost of travel to meetings to discuss the transaction), opportunity costs (e.g. time invested in exploring incentive options that would otherwise have generated an income), and non-monetary costs (e.g. time spent managing business affairs instead of recreation). A summary of the likely range of transaction costs in agri-environment schemes, drawing in particular on Coggan et al. (2010), is set out in Table 16.1.

Table 16.1: Likely transaction costs encountered in agri-environment schemes.

Type of transaction cost	When it is incurred	Scheme proponent and administrator costs	Payment recipient costs
Information about the problem	Well before the scheme has been decided upon (even many years before).	Identifying, collecting, and analysing data about the problem and potential solutions.	Participation in problem scoping and providing information.
Scheme selection and development	Months to years prior to scheme implementation.	Examining policy options and consulting with stakeholders.	Participation in consultation, lobbying for preferred option.
Establishment	Immediately prior to landholder engagement.	Staff training, equipment, systems set-up, advertise and promote.	Gathering information about scheme, and preparation to engage.
Implementation (including repeated implementation)	Initial selection and contracting phase — repeated as needed.	Engage with and process participants, negotiate contracts, etc.	Engage with scheme, prepare proposals, negotiate contracts, etc.
Scheme management	Ongoing scheme management such as making payments, basic reporting, and so on.	Make payments, record keeping, engagement as required.	Reporting, record keeping.
Landholder monitoring and compliance	After contracting — auditing and any enforcement required.	Auditing and verifying reporting, any compliance activities.	Defence of compliance activities, additional reporting, etc.
Ecological monitoring and evaluation	Before, during and after scheme (length depends on ecological response time).	Data collection and evaluation of ecological outcomes (relative to problem formulation).	Likely to be relatively low.
Scheme evaluation and improvement	During and after contract completion.	Incurred in analysis of effectiveness, making and implementing recommendations.	Lobbying for scheme changes, etc.

These costs will vary from scheme to scheme depending on:

- The type of transaction (e.g. simple versus complex, once-off or long-term) which is linked to the type of scheme. It will also depend on transaction uniqueness, frequency, and uncertainty in various elements, including ease of observing actions and outcomes, time-lags to environmental effects.

- Who is involved in the transaction — especially their prior experience, opportunism or extent of self-interested strategic behaviour.

- Other influences including previous and other policy and procedures, trust in administrators, comfort with policy principles and social connectedness (Coggan et al. 2010).

Two recent themes in the transaction cost literature have refined our understanding of transaction costs in agri-environment schemes. Firstly, transaction costs have a dynamic element through time (which partly covers our earlier point about when transaction costs can be considered sunk and thus irrelevant to informing current decisions) as the effectiveness of instruments through time and under different demands will change with experience, technology, and other factors (Arrow 1962; Falconer et al. 2001; Fang et al. 2005; McCann et al. 2005). This is particularly relevant to trading schemes, but could also apply to impacts of technological change on, for example, monitoring in grant schemes.

Secondly, a broader emphasis has arisen towards the role of transaction costs in evaluating overall policy efficiency, rather than on measuring transaction costs in isolation (see Pannell et al. 2013, for example). There are a number of very good studies either stepping through the process of measuring the transaction costs of agri-environmental and broader environmental policy (see McCann et al. 2005; Kuperan et al. 2008) or measuring transaction costs from actual schemes (see Falconer and Saunders 2002; Mettepenningen and Van Huylenbroeck 2009; Ofei-Mensah and Bennett 2013).

Transaction costs are not trivial

A number of researchers have directed their attention toward the empirical assessment of transaction costs across a wide range of settings. Estimates of transaction costs to scheme administrators and proponents have ranged from <1–100 per cent of the payments made to scheme participants. For instance, studies carried out in the US show that public transaction costs represent a substantial part of total costs incurred in designing a policy objective, with a magnitude ranging from 8 per cent of the water purchase cost (Howitt 1994) to 38 per cent of the agricultural assistance program (McCann et al. 2005). Public administration transaction costs of agri-environmental schemes across Europe was initially 102 per cent of payments to landholders (1992/93) but declined over time to 18 per cent (1998/99) (Falconer et al. 2001). Mettepenningen and Van Huylenbroeck (2009) explicitly look at private transaction costs of an agri-environmental scheme and report that, on average, these are 15 per cent of the total cost of the policy.

Recent Australian estimates have indicated that the average transaction costs amount to nearly $8,400 for Reef Rescue participants (average total transaction cost per farm was 38 per cent of average funding provided) (Coggan et al. 2014). On the proponent and administrator side of the equation, Binney et al. (2010) indicate that the costs to the Australian Government of the Environmental Stewardship Program was around 10 per cent, and the (Tasmanian) Forest Conservation Fund was around 11 per cent of total scheme costs, although these exclude some elements of investigation, design, and establishment. Our own experience with one regional Victorian catchment management authority with substantial experience in agri-environment scheme delivery identified ongoing transaction costs (i.e. only implementation, scheme management, and some elements of monitoring and compliance) of around 10 per cent of total costs for a tendering program, and 95 per cent of program costs for a parallel grant scheme. Differences resulted from the grant scheme funding smaller projects, requiring a minimum cost-share from landholders, and apparently requiring additional recruitment effort into a less flexible program.

So we can see that transaction costs are likely to vary across applications and the type of scheme implemented. In Table 16.2, we set out some of the likely differences in costs across different types of agri-environment schemes, namely grants (rule-based allocation of funds, often with or without consideration of cost-effectiveness), tendering or reverse auction approaches (funds allocated competitively to landholder applicants based on relative cost-effectiveness), and offset schemes (usually negotiated contracts with landholders). We note that indicating the per unit costs is difficult because transaction costs are made up of fixed and variable costs. Fixed costs are not strongly influenced by the amount of payments or participants, while variable costs are directly related to participation.

Table 16.2: Likely differences in transaction costs between agri-environmental payment approaches.

Transaction cost	Grants	Conservation tenders	Offset payments
Information about the problem	Costs unlikely to differ across schemes, although there will likely be additional lobbying and consultation costs in two-sided markets, such as offsets.		
Scheme selection and development	Usually can be adapted from existing approach (likely to be less reliance on metric for assessment).	Greater costs in bid assessment and possibly in differential contracts.	Design of more rigorous assessments on both market sides and potentially a trading mechanism.
Establishment	Existing processes with specialised advertising and engagement.	Specialised advertising and engagement, new systems to receive and rank bids. Often more detailed training for field assessment.	Detailed registry often required in addition to measurement systems. Sometimes trust arrangements for funds.
Implementation (including repeated implementation)	Site visits remain expensive. Covenants uncommon.	Any field assessment will remain relatively expensive. More detailed contract, engagement. Covenants incur substantial time and legal costs where relevant.	Unlikely to differ substantially from tenders, but continue to be incurred whilst tenders tend to be one-off events. Covenants almost universally required and incur substantial transaction costs.
Scheme management	Depends on scheme design, but usually low.	Landholders may need to submit detailed annual reports. May be ongoing payments.	Ongoing cost of finding and negotiating with offset supplier, making payments and conducting reporting.
Landholder monitoring and compliance	Often little or no landholder self-monitoring.	Usually at least some landholder self-monitoring.	Usually at least some offset provider ongoing self-monitoring required.
	Compliance monitoring is low across all schemes in practice but should logically be higher for more complex and ongoing schemes.		
Ecological evaluation and program evaluation and improvement	Highly dependent on whether the scheme is formally evaluated or repeated. Will usually be higher for a trading mechanism because it is explicitly ongoing but so little is done in practice that costs unlikely to differ.		

So which costs are likely to matter most? Available data mostly focuses on costs from the establishment stage onwards. For example, the National Market Based Instruments Pilots program (BDA Group 2009) and Binney et al. (2010) focus on costs to government and natural resource management (NRM) groups once a particular approach has been selected. Most of these establishment costs are likely to be fixed (though investment in human skills can be easily lost). The major transaction costs to government are likely to be implementation costs, particularly where site visits and heterogeneous or customised contracts are required. While transaction costs such as these are often unavoidable, some measures can be taken to reduce the cost. Retaining corporate knowledge on processes and data collection methodologies for site visits is one solution. Targeting schemes to critical landholders where customised contracts are required is another. Metric design is often thought to be expensive, but was less than 1 per cent of the environmental stewardship program budget (Binney et al. 2010). Good data storage and use of corporate knowledge could have been contributing factors here. On the private side, transaction costs appear to be driven by complexity of both the scheme and the individual interaction required more than scheme type, although there are few studies that actively examine the differential aspects of transaction costs (see Chapter 21 for some insights on conservation tenders, for example). Timely, clear and consistent communication of scheme requirements by the administrator, along with easy access to information for private parties, can significantly reduce private transaction costs (Coggan et al. 2013).

Conclusions

Transaction costs for agri-environment schemes are the cost of time and effort, as well as direct expenditure incurred in scheme investigation, design, implementation, management and administration, and monitoring and evaluation. The scale and distributional burden of transaction costs should be carefully considered alongside other costs and benefits from agri-environment schemes, as an essential element of understanding whether government policy is efficient, or at least cost-effective. Of course, in some settings, some costs are sunk and should not be considered in an analysis of the comparative efficiency of different schemes. Hence, it is important for scheme proponents and designers to identify which are likely to be the most important transaction costs, who bears them, and how they impact on different elements of scheme design or implementation.

The purpose of considering transaction costs is not necessarily to minimise or reduce them. Instead, a focus on efficient program design may require increased transaction costs in order to focus more closely on delivering the desired outcome. That is, different program designs will have different implications for transaction costs and overall efficiency. Despite the increasing number of transaction costs analyses, none focus on the potential for efficiency dividends to be achieved from higher transaction costs, and this would seem to be a particularly useful area to explore given the emphasis on minimising delivery overheads. Delivery and implementation of agri-environment schemes can benefit from a closer focus on transaction costs.

References

Arrow, K.J. (1962) 'Economic welfare and the allocation of resources for invention', *The rate and direction of inventive activity: Economic and social factors* (ed. R.R. Nelson), Princeton, Princeton University Press.

BDA Group (2009) *Final report of the National Market Based Instruments Pilot Program*, prepared for the Department of Agriculture, Fisheries and Forestry.

Binney, J., K. Whiteoak and G. Tunny (2010) *Review of the Environmental Stewardship Program*, Marsden Jacob and Associates, Available at: www.nrm.gov.au/resources/publications/stewardship/esp-review. html.

Coggan, A., E. Buitelaar, S.M. Whitten and J. Bennett (2013) 'Factors that influence transaction costs in development offsets: Who bears what and why?', *Ecological Economics* 88: 222–31.

Coggan, A., M. van Grieken, A. Boullier and X. Jardi (2014) 'Private transaction costs of participation in water quality improvement programs for Australia's Great Barrier Reef: Extent, causes and policy implications', *Australian Journal of Agricultural and Resource Economics* 59(4): 499–517.

Coggan, A., S.M. Whitten and J. Bennett (2010) 'Influences of transaction costs in environmental policy', *Ecological Economics* 69: 1777–84.

Falconer, K. and C. Saunders (2002) 'Transaction costs for SSSIs and policy design', *Land Use Policy* 19:157–166.

Falconer, K., P. Dupraz, M. Whitby (2001) 'An investigation of policy administrative costs using panel data for the English environmentally sensitive areas', *Journal of Agricultural Economics* 52: 83–103.

Fang, F., K.W. Easter and P.L. Brezonik (2005) 'Point-non-point source water quality trading: A case study in the Minnesota River Basin', *Journal of American Water Resources Association* 41: 645–58.

Howitt, R.E. (1994) 'Empirical analysis of water market institutions: The 1991 California water market', *Resource and Energy Economics* 16, 357–71.

Kuperan, K., N.M.R. Abdullah, R.S. Pomeroy, E.L. Genio and A.M. Salamanca (2008) 'Measuring transaction costs of fisheries co-management', *Coastal Management* 36: 225–40.

McCann, L., B. Colby, K.W. Easter, A. Kasterine and K.V. Kuperan (2005) 'Transaction cost measurement for evaluating environmental policies', *Ecological Economics* 52: 527–42.

Mettepenningen, E. and C. Van Huylenbroeck (2009) 'Factors influencing private transaction costs related to agri-environmental schemes in Europe', *Multifunctional rural land management: Economics and policies* (eds F. Bruwer and M. van der Heide), Earthscan, London, pp. 145–68.

North, D.C. (1990) *Institutions, institutional change and economic performance*, Cambridge University Press, Cambridge.

Ofei-Mensah, A. and J. Bennett (2013) 'Transaction costs of alternative greenhouse gas policies in Australian transport energy sector', *Ecological Economics* 88: 214–21.

Ostrom, E. (1990) *Governing the commons: The evolution of institutions for collective action*, Cambridge, Cambridge University Press.

Pannell, D.J., A.M. Roberts, G. Park and J. Alexander (2013) 'Improving environmental decisions: A transaction-costs story', *Ecological Economics* 88: 244–52.

Reeson, A.F., L.C. Rodriguez, S.M. Whitten, et al. (2011) 'Adapting auctions for the provision of ecosystem services at the landscape scale', *Ecological Economics* 70: 1621–7.

Williamson, O.E. (1998) 'Transaction cost economics: How it works; where it is headed', *De Economist* 146(1): 23–58.

17

What a difference a metric makes: Strong (and weak) metrics for agri-environment schemes

Fiona Gibson and David Pannell

Key lessons

- A range of metrics are used to evaluate and prioritise projects within agri-environment schemes.
- The way the metric is calculated, and the choice of variables included, are important decisions in the evaluation process.
- When funds are scarce, the quality of the metric is important.
- Errors in metric design are readily avoidable.
- It is more important to ensure that high-quality decision metrics are used than to invest in improving the quality of information about projects.

Good decision-making in agri-environment schemes is information-intensive. Environmental managers usually collect and weigh up information on landscape characteristics, ecological responses, human behaviour, and project risk. This information feeds into their decision-making. Environmental managers usually put a lot of effort into collecting this information, but often take a rough-and-ready approach to combining it into a form that is useful for decision-making. For example, for investments made in the National Action Plan, Pannell and Roberts (2010) commented: 'The processes used by Catchment Management Organisations generally did not involve comprehensive systematic analysis of investment options or project design options.'

Figure 17.1: Failed revegetation project in northern Victoria. Effective decision metrics can help to identify and prioritise projects that are more likely to succeed.
Source: Photo by David Freudenberger.

Does this matter? Does it make a difference to environmental outcomes to use a theoretically sound decision metric, compared with a weak decision metric? That was the question we set out to answer by comparing environmental outcomes generated by these two approaches.

What we found, in short, was that it does matter which decision metric you use. Indeed, it can make an enormous difference. As a consequence, many decision metrics used by environmental managers result in us missing out on very large environmental benefits.

What's in a metric?

What is a decision metric and why are they so important? Around the world, billions of dollars worth of public funds are allocated to environmental projects each year. These funds are scarce, relative to the amount needed to support all possible environmental projects, so prioritisation is essential. This means some projects are determined to be more valuable than others and will receive funding whereas the less valuable projects miss out.

A common approach used by environmental managers to score the projects they have to choose between is to define a set of variables believed to correlate with projects' benefits and costs, and combine them into a formula or metric so that projects can be compared. Numerical values or scores are assigned to each potential project and these scores are used to rank the projects. For example, the Conservation Reserve Program in the United States combined measures of wildlife benefit, water quality benefit, erosion risk, enduring benefit, air quality benefit, priority area, and cost to evaluate program investments (Hajkowicz et al. 2009).

Of course, there are many different ways the various benefits and costs of a project could be combined, and there are thousands of different decision metrics in practice around the world. Unfortunately, many if not most of these decision metrics have problems in the way they determine the value of the project. Indeed, our analysis showed that the performance of many of these metrics is not much better than choosing projects at random. If that's the case, there's little point in wasting your time on using these metrics — which take time and money to generate — because you may as well simply draw projects out of a hat. Commonly used decision metrics have a range of weaknesses, including adding variables that should be multiplied, omitting important variables related to environmental benefits, omitting project costs, or subtracting costs rather than dividing by them (see Box 17.1).

But what do these weaknesses add up to in terms of lost value? Surprisingly, few have undertaken such analysis (see Joseph et al. 2009). We estimated the environmental losses resulting from each of these weaknesses.

Box 17.1: Divide benefit by the cost — don't subtract.

The first principle of creating a strong metric is understanding that it reflects a measure of project benefits divided by a measure of project costs. Economists call this metric a benefit–cost ratio (BCR).

$$BCR = \frac{B}{C}$$

There are plenty of project ranking metrics in actual use that don't do this. Some subtract costs instead of dividing them, and some (remarkably) ignore costs entirely. These are mistakes that are costly to the environment.

To illustrate this, consider the following three hypothetical projects, with the indicated benefits (B) and costs (C). Because the budget is limited, the first project we should choose is the one with the highest benefits per unit cost (the highest BCR), which is project one. But if we rank according to B-C (i.e. benefit minus the cost), the top ranked project seems to be project two, while ranking according to just B (i.e. benefit ignoring costs altogether) tells us that project three is best.

The loss of environmental values from using the wrong metric (i.e. ranking according to B-C or B) depends on how tight the budget is. Assuming that the budget is enough to fund 10 per cent of projects, the loss of environmental benefits is 12 per cent for B-C, and 19 per cent for B (based on simulating 1,000 funding rounds with 100 potential projects in each).

In other words, fixing up the formula is like increasing the program budget by 14 per cent or 23 per cent. It's much easier to fix the formula than to increase the budget.

Project	B	C	BCR	B-C	Rank (BCR)	Rank (B-C)	Rank (B)
1	5	1	5	4	1	2	3
2	7	2	3.5	5	2	1	2
3	8	7	1.1	1	3	3	1

The attributes of a robust metric

Pannell (2013) described the requirements for a theoretically sound and practical decision metric for ranking environmental projects. He recommends:

$$BCR = \frac{V \times W \times A \times (1 - R)/(1 + r)^L}{C}$$

where BCR stands for benefit–cost ratio, and benefits depend on the value (V) of the environmental assets; the likely adoption of new practices or behaviours (A); the effectiveness of the new practices at increasing environmental values (W); the risk of project failure (R); the time lag until benefits occur (L); and the discount rate (r). Benefits are divided by costs (C) to derive the BCR, with higher BCRs demonstrating a more cost-efficient project. All of the benefit-related variables are multiplied, not weighted and added, for reasons explained by Pannell (2013). We obtained distributions for each of these variables from a database of 129 projects that have been evaluated using INFFER (the Investment Framework for Environmental Resources — see Chapter 18).

Essentially, our analysis involved evaluating and ranking projects using Pannell's metric (given above) and an alternate metric with one or more weaknesses included — for example, $BCR = V{\times}W{\times}A{\times}(1-R)/(1+r)^{L}$, which omits costs ($C$). By comparing the two results, we estimated the overall loss of environmental values from selecting relatively weak projects using the alternative metric. We tested the metrics for different program budget levels — from 2.5 per cent to 40 per cent of the budget required to fund all the projects. Altogether, the analysis simulated 27 million projects being considered in 270,000 project-prioritisation decisions.

What's lost?

Using weak metrics makes an enormous difference: the wrong projects get funded, resulting in big losses of environmental values. Where funding is tight (as it almost always is) we found that poor metrics resulted in environmental losses of up to 80 per cent — not much better than completely random, uninformed project selection.

The most costly errors were found to be omitting information about environmental values, project costs, or the effectiveness of management actions. Using a weighted-additive decision metric for variables that should be multiplied is another costly error commonly made in real-world decision metrics (e.g. adding cost in the Conservation Reserve Program's environmental benefits index). We found that omitting information about project costs or the effectiveness of management actions, or using a weighted-additive decision metric (that should be multiplied) can reduce potential environmental benefits by 30 to 50 per cent. Think about how hard it would be to double your budget (achieve a bigger slice of the funding pie), and yet that could be achieved in effect in many cases by simply strengthening the decision metric being used.

What about the quality of the information?

Of course, it's not just the structure of the metric calculation that could be a weakness in the prioritisation. The quality of the information going into the calculation is also a factor (see Anderson et al. 1977 for a description of the standard theoretical framework for calculating the value of information to an application in agriculture). We looked at the environmental losses resulting from use of poor-quality information, such as inaccurate cost data, in the decision metric. We compared results from prioritising projects based on perfect information and uncertain information.

Naturally, poorer quality information about projects results in some relatively weak projects being selected for funding. Surprisingly, however, we found that the quality of the decision metric makes a much bigger difference to environmental outcomes than the quality of the information used within it.

If a very poor metric is used, then the benefits of improving data quality from high uncertainty to perfect information are remarkably low: 3 to 6 per cent. Improving information quality (e.g. by collecting more or different types of data) only produces benefits greater than 10 per cent if a reasonably good decision metric is used, and even then only if the available budget is tight.

That's an amazing finding which suggests environmental managers (and policymakers) should be more concerned in the first instance about how they calculate a decision metric rather than funding the acquisition of higher quality (and inevitably much more expensive) information to feed into that metric (see chapters 20 and 21).

Does it really matter?

Our results show that relatively simple improvements to metrics used for environmental decision-making can make a big difference to the environmental benefits generated by funded projects. Environmental budgets are usually small, relative to the problems faced, so good decision metrics are crucial.

It does really matter which decision metric you use. Another way of thinking about this is by considering how much effort people put into increasing environmental budgets. Of course, getting a bigger slice of the budget pie will help in achieving environmental outcomes. However, this analysis suggests that efforts to improve environmental decision processes may be even more beneficial than equivalent efforts devoted to increasing the total environmental budget. With less funding available for agri-environment schemes, the design of high-quality project selection metrics is critical.

References

Anderson, J.R., J.L. Dillon and J.B. Hardaker (1977) *Agricultural Decision Analysis*, Iowa State University Press, Ames.

Hajkowicz, S., K. Collins and A. Cattaneo (2009) 'Review of agri-environment indexes and stewardship payments', *Environmental Management* 43: 221–36.

Joseph, L.N., R.F. Maloney and H.P. Possingham (2009) 'Optimal allocation of resources among threatened species: A project prioritization protocol', *Conservation Biology* 23: 328–38.

Pannell, D.J. (2013) *Ranking environmental projects*, Working Paper 1312, School of Agricultural and Resource Economics, University of Western Australia.

Pannell, D.J. and F.L. Gibson (2014) 'Testing metrics to prioritise environmental projects', Working Paper 1401, School of Agricultural and Resource Economics, University of Western Australia.

Pannell, D. and A.M. Roberts (2010) 'Australia's National Action Plan for Salinity and Water Quality: a retrospective assessment', *Australian Journal of Agricultural and Resource Economics* 54: 437–56.

Pannell, D., A.M. Roberts, G. Park and J. Alexander (2013) 'Designing a practical and rigorous framework for comprehensive evaluation and prioritisation of environmental projects', *Wildlife Research* 40: 126–33. Available at: dx.doi.org/10.1071/WR12072.

18

Public benefits, private benefits, and the choice of policy tool for land-use change

David Pannell

Key lessons

- The selection of the best policy tool or delivery mechanism for an agri-environmental project depends crucially on the levels of public net benefits and private net benefits generated by the project.

- A framework is presented that recommends a policy mechanism from one of five categories: (a) positive incentives; (b) negative incentives; (c) extension (technology transfer, education, communication, demonstrations, support for community network); (d) technology development; and (e) no action.

- Private net benefits (which drive landholder behaviour) are just as important as public net benefits (e.g. for the environment) when selecting the policy mechanism.

- Australian programs tend to rely too much on extension and too little on positive incentives and technology development.

- Program managers can use the framework provided here to better match policy mechanisms to particular projects and programs.

Figure 18.1: Conservation actions can generate private benefits — for example, addressing land degradation issues, such as erosion, that impact on agricultural production.
Source: Photo by Dean Ansell.

Agri-environmental programs around the world have been created to attempt to encourage changes in land management on privately owned lands in order to enhance environmental conservation or natural resource management (NRM). To encourage change, these programs use a range of policy mechanisms that can be categorised as (a) positive incentives (financial or regulatory instruments to encourage change), (b) negative incentives (financial or regulatory instruments to inhibit change), (c) extension (technology transfer, education, communication, demonstrations, support for community network), and (d) technology change (development of improved land management options, such as through strategic R&D, participatory R&D with landholders, or provision of infrastructure to support a new management option). A fifth option available to governments is (e) no action, which can be appropriate if the cost of achieving a desired change is so large that it outweighs the resulting benefits, or if the benefits are expected to occur without government intervention.

In practice, the choice among these possible policy mechanisms is often not very sophisticated. Programs primarily tend to rely on a small number of mechanisms, sometimes as few as one. The choice among these mechanisms depends on the levels of public net benefits and private net benefits from the land-use changes being proposed. Private net benefits refer to benefits minus costs accruing to the private land manager as a result of the proposed changes in land management. Private net benefits are the main driver of adoption of new innovations, and depend on a wide range of factors, not just profits (Pannell et al. 2006).

Public net benefits means benefits minus costs accruing to everyone other than the private land manager. Defining these terms in these ways is helpful because the private net benefit dimension provides insight into the behaviour of the landholder, while the public net benefit dimension relates to the effects on everyone else that flow from the landholder's behaviour.

The Public: Private Benefits Framework

In this chapter, I present the Public: Private Benefits Framework, a simple tool that helps to identify which type of policy instrument is most suitable for a particular agri-environmental project or program, based on the levels of public and private net benefits that are likely to result from the land-use change (Pannell 2008). It is useful in cases where environmental managers wish to influence the management of private lands to promote the conservation of natural resources or the environment. It is based on levels of public and private net benefits of changing land management, and a set of simple rules. It provides a powerful tool for targeting environmental investments to high-payoff projects, and for selecting policy mechanisms that are most likely to be cost effective.

The starting point for the framework is the recognition that agri-environmental managers can invest in a range of projects involving changes in land management or land use on private land, and that the available options vary widely in the levels of public and private net benefits they generate, potentially including negative net benefits. The aim is to identify which policy mechanisms are likely to be suitable for each potential project.

To select policy mechanisms, the following set of rules is proposed, leading to Figure 18.2.

1. Do not use positive incentives for land-use change unless public net benefits of change are positive.

2. Do not use positive incentives if landholders would adopt land-use changes without those incentives.

3. Do not use positive incentives if private net costs outweigh public net benefits.

4. Do not use extension unless the change being advocated would generate positive private net benefits. In other words, the practice should be sufficiently attractive to landholders for it to be adoptable once the extension program ceases.

5. Do not use extension where a change would generate negative net public benefits. Note that rules four and five are referring to cases where extension is used as the main tool to achieve land-use change. Extension could also be used to support any of the other policy mechanisms, playing a supporting role rather than being the main tool.

6. If private net benefits are negative (but not overly negative), consider technology development to create improved (environmentally beneficial) land management options that can be made adoptable with or without positive incentives (Pannell 2009).

7. If private net benefits outweigh public net costs (such that the project would have negative net benefits overall — in other words, a net cost), the land-use changes could be accepted if they occur, implying no action, or they could be penalised at an appropriate level, but not prohibited. The latter approach uses a pricing mechanism to force landholders to consider the negative consequences of their actions. This allows them to weigh up whether their benefits exceed those negative consequences, thus making prohibition unnecessary.

8. If public net costs outweigh private net benefits, use negative incentives.

9. If public net benefits and private net benefits are both negative, no action is necessary. Adverse practices are unlikely to be adopted.

10. In all cases, the suggested action needs to be weighed up against a strategy of no action.

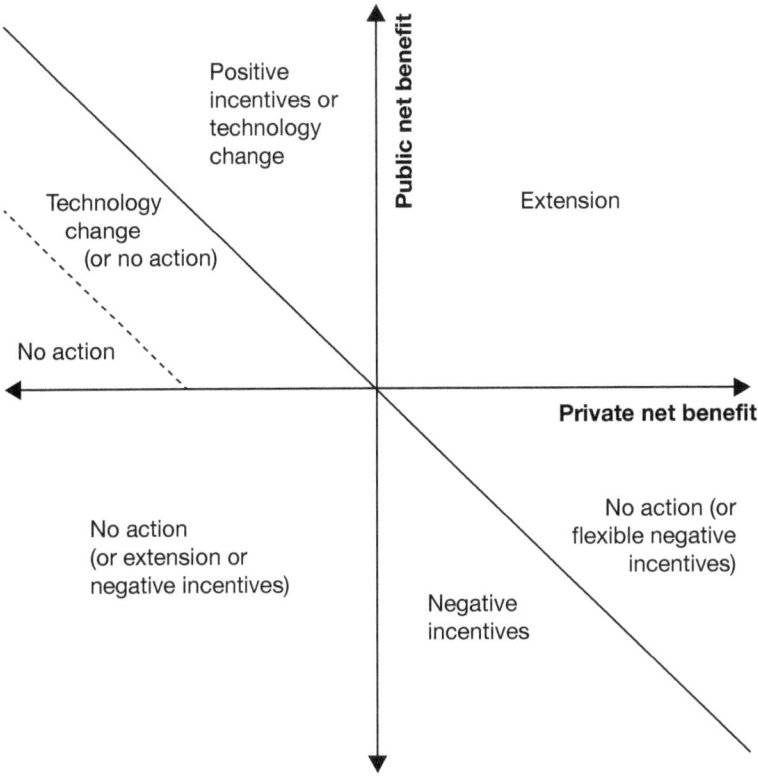

Figure 18.2: Recommended efficient policy mechanisms based on a simple set of rules.
Sources: Author's research.

For any given project, the levels of public net benefits and private net benefits, relative to current practice (which is represented by the zero–zero point in the centre), are estimated and plotted on the graph. Depending on the location of the project on the graph, the appropriate policy response is indicated. For advice on how to estimate public net benefits and private net benefits, see my 'Frequently Asked Questions' page for the framework: dpannell.fnas.uwa.edu.au/ppf-faq.htm.

Note that the zero–zero point remains the current practice no matter how good or bad that practice is with regard to public net benefits. The relevant questions addressed by the framework are (a) whether it is possible and worthwhile to do better than current practice, (b) whether it is worthwhile stopping or discouraging landholders from switching from their current practice to something more environmentally damaging, and (c) if so, how? These remain relevant questions whether the current practice is highly damaging or highly beneficial to the environment.

When estimating net benefits, if there are time lags until the realisation of costs or benefits, these should be discounted using standard discounting methods. The public and private net benefits that are graphed would thus be present values (Pannell and Schilizzi 2006).

This is quite a simple framework, but it is a good start. It significantly narrows down the range of policy tools that environmental managers should be considering depending on public and private net benefits in a particular situation. We can make it more sophisticated in various ways, including by allowing for time lags until adoption, learning costs involved in land-use change, the fact that extension reduces but does not eliminate lags to adoption, the transaction costs of extension, and through requiring higher levels of selectivity (a higher benefit–cost ratio) than just covering costs. Figure 18.3 allows for these complexities, and requires a benefit–cost ratio of at least two.

In broad terms, the framework advocates the use of:

- positive incentives if the public net benefits of land-use change are high, and the private net benefits are not too negative;
- extension if the public net benefits of land-use change are high, and the private net benefits are moderate;
- no action if private net benefits are positive and public net benefits are not sufficiently high;
- no action if private net benefits are greater than public net costs;
- negative incentives if private net benefits are less than public net costs;
- no action if public net benefits and private net benefits are both negative; and
- technology development if private net benefits are low-to-moderately negative and public net benefits are positive (Pannell 2009).

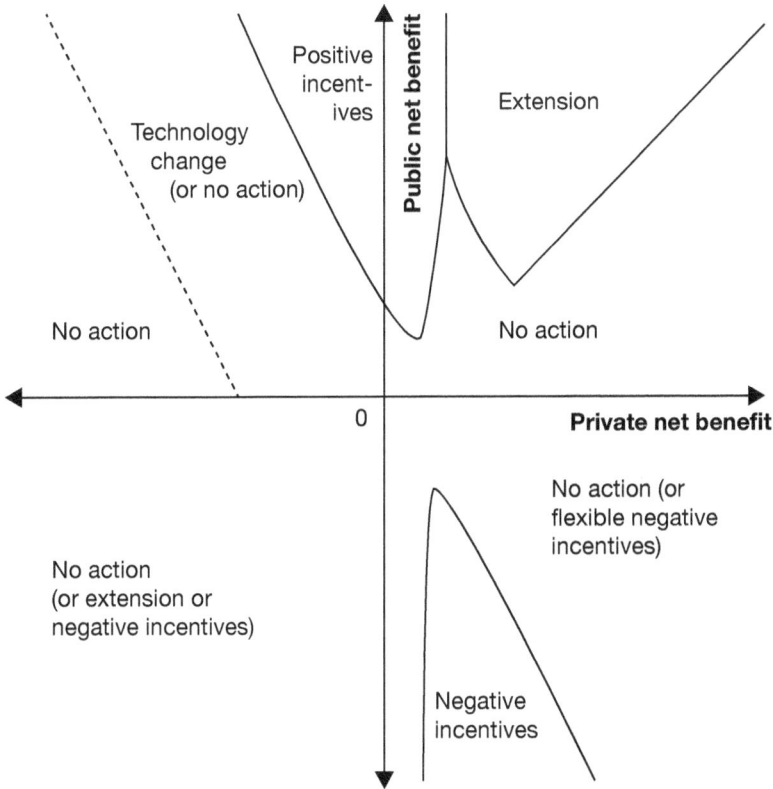

Figure 18.3: Efficient policy mechanisms for encouraging land use on private land, refined to account for various complexities described in the text. A smaller number of projects would qualify for incentives or extension in this more targeted approach, relative to Figure 18.2.

Source: Author's research.

Past Australian agri-environmental programs have often been inconsistent with these insights. For example, the National Action Plan for Salinity and Water Quality relied largely on extension and small temporary incentive payments (which are essentially a form of extension), but promoted practices that, in most cases, were not adoptable at the required scale (Pannell and Roberts 2010). For many of the land management strategies that were promoted, public net benefits were low (Graham et al. 2010) and private net benefits were highly negative (Kingwell et al. 2003), such that the projects would have fallen in to the 'no action' or 'technology change' sections of the framework's top left quadrant. The program made almost no investment in development of improved technologies for salinity management, even though this was the strategy most likely to succeed in many areas.

The Public: Private Benefits Framework is embedded within INFFER (the Investment Framework for Environmental Resources) (Pannell et al. 2012). When a project is evaluated using INFFER's Project Assessment Form, the location of that project on the Public: Private graph is provided as an output. The framework can be modified to suit particular contexts or target groups. For example, Pannell and Wilkinson (2009) adjusted it for lifestyle landholders (also known as hobby farmers). Their adjustments involved increasing the transaction costs per hectare (reflecting the small sizes of these properties) and increasing the private net benefits from environmental actions.

The framework highlights the importance of targeting funds in environmental programs to selected areas, based on the levels of public and private net benefits. Environmental managers tend to be focused on the level of public benefits when selecting their investments, but often pay inadequate attention to the level of private net benefits, which, perhaps surprisingly, turns out to be even more important as a driver of policy decisions.

It is worth noting that the rules underlying the framework are based on an objective of efficiency (biggest environmental benefit per dollar spent). In practice, governments often also pursue other objectives, ranging from perceived equity to raw political motives. I hope that by improving the understanding of what an efficient policy would look like, this framework can make it easier for that objective to be pursued.

Acknowledgements

Thanks to the ARC Centre of Excellence for Environmental Decisions and the Australian Government's National Environmental Research Program Environmental Decisions hub for funding support.

References

Graham, T., D.J. Pannell and B. White (2010) 'Determining the net-benefits from government intervention for dryland salinity: A breakeven analysis', *Australasian Journal of Environmental Management* 17(2): 112–24.

Kingwell, R., S. Hajkowicz, I. Young, et al. (2003) *Economic evaluation of salinity management options in cropping regions of Australia*, Grains Research and Development Corporation, Canberra.

Pannell, D.J. (2008) 'Public benefits, private benefits, and policy intervention for land-use change for environmental benefits', *Land Economics* 84(2): 225–40. Available at: dpannell.fnas.uwa.edu.au/ppf.htm.

Pannell, D.J. (2009) 'Technology change as a policy response to promote changes in land management for environmental benefits', *Agricultural Economics* 40(1): 95–102.

Pannell, D.J., G.R. Marshall, N. Barr, et al. (2006) 'Understanding and promoting adoption of conservation practices by rural landholders', *Australian Journal of Experimental Agriculture* 46(11): 1407–24.

Pannell, D.J. and A.M. Roberts (2010) 'The National Action Plan for Salinity and Water Quality: A retrospective assessment', *Australian Journal of Agricultural and Resource Economics* 54(4): 437–56.

Pannell, D.J., A.M. Roberts, G. Park, et al. (2012) 'Integrated assessment of public investment in land-use change to protect environmental assets in Australia', *Land Use Policy* 29(2): 377–87.

Pannell, D.J. and S. Schilizzi (eds) (2006) *Economics and the future: Time and discounting in private and public decision making*, Edward Elgar, Cheltenham, UK and Northampton, MA, USA.

Pannell, D.J. and R. Wilkinson (2009) 'Policy mechanism choice for environmental management by non-commercial "lifestyle" rural landholders', *Ecological Economics* 68: 2679–87.

19

Controls and counterfactual information in agro-ecological investment

David Duncan and Paul Reich

Key lessons

- Appropriate contrasts, such as controls and counterfactual data, are fundamental to sound interpretation of the effectiveness of agri-environment schemes.
- Such contrasts are rarely included in evaluations of Australian agri-environment schemes for a range of reasons, including logistical constraints.
- Different kinds of contrasts exist that permit different kinds of inference about program effectiveness.
- Effective evaluation incorporating sampling counterfactual data need not cost more than is currently expended on monitoring and evaluation.
- Every scheme should explicitly include counterfactual thinking in evaluation plans, even if there is no intention to monitor.

Figure 19.1: A riparian zone near Euroa, Victoria, that has been fenced and replanted with a mix of native species. Evaluating the benefits of projects such as these requires an understanding of what would have happened in the absence of the project — the counterfactual.

Source: Photo by Paul Reich.

Introduction

Despite the large amounts of money invested in agri-environment schemes in Australia (Hajkowicz 2009), there remains high uncertainty about the magnitude of expected environmental benefit. Sophisticated approaches for dealing with uncertainty are now routinely adopted in systematic conservation planning processes (e.g. prioritisation, optimisation, Sarkar et al. 2006). Unfortunately, these advances have not been matched in the evaluation, learning, and improvement part of the decision cycle, where conservation and environmental management lag behind other complex domains such as social policy and medicine (Stem et al. 2005; Ferraro 2009; Field et al. 2007).

Reporting of management performance (activity and outputs, *sensu* Mascia et al. 2014) in agri-environment schemes has itself been patchy, but direct demonstrations of the impact of intervention, that is, the difference in change between intervention and non-intervention sites, are exceedingly rare (Margoluis et al. 2009; Ferraro and Pattanayak 2006; but see Hale et al. 2011; Lindenmayer et al. 2012). As we discuss in this chapter, this so-called counterfactual evidence (derived from control sites without intervention) is fundamentally important for meaningful evaluation of agri-environmental investment schemes. We outline the difficulties that investment in environmental change poses for management experiments, and suggest positive ways of addressing those difficulties.

Our focus is the type of evaluation questions identified by Mascia et al. (2014) as 'impact evaluation' where the interpretation logically demands a counterfactual contrast (Ferraro 2009). For example, what is the impact of livestock exclusion from remnant vegetation on the abundance of sensitive native herbs?

In most cases, the effectiveness of work funded by agri-environment schemes is evaluated using post-hoc, space-for-time substitution surveys (e.g. Prober et al. 2011; Read et al. 2011). To accept the implied effect, we assume that the sites were equivalent at some point in the past and that the difference is due to the funded intervention, but we cannot be certain of this. Particularly in agricultural settings, many confounding factors exist that could inflate or obscure the effects of the intervention. Also, we ignore the possibility of influence of interactions between, for example, climatic regime and the intervention. More powerful conclusions about the effectiveness of changed management can be made when data are collected through time, for the period of the intervention.

Responses to large-scale interventions have been evaluated where sufficient long-time series data were available to allow change point analysis (i.e. the detection of a change in a time series, Box and Tiao 1975; Stewart-Oaten and Bence 2001; Thomson et al. 2010; Stoffels and Weatherman 2014). These examples have no direct counterfactual sampling option because the subject or events cannot be replicated meaningfully (e.g. matching an entire river, estuary, city, etc.). Instead the interpretation relies on consideration of whether the observed change could have been generated independently of the intervention. This is one example of model-based counterfactual inference, where an expected alternative under no intervention may be credibly argued with reference to weight of data accumulated prior to the intervention.

By contrast, the investments in agri-environmental schemes are usually conceptually and practically replicable, although we acknowledge that important constraints exist. Time series analysis based on intervention sites are usually impossible because the locations where interventions will occur are rarely known for long enough in advance to enable adequate data collection, and the interventions themselves are relatively short term. This means that direct counterfactual inference must come from simultaneous sampling of control and intervention sites.

Why are controls needed to estimate the impact of agri-environmental schemes?

By making payments, agri-investment schemes seek to change the status quo. The basic assumption is that those sites or landscapes where investment (intervention) is made will have a different future to sites where no intervention occurs. When change is estimated only from intervention sites, there is an implicit assumption that non-funded sites will not change (e.g. Figure 19.2a). While this is one plausible scenario, there are others that should be considered. Counterfactual information from control sites allows us to weigh rival interpretations of the outcomes of interventions (Ferraro 2009).

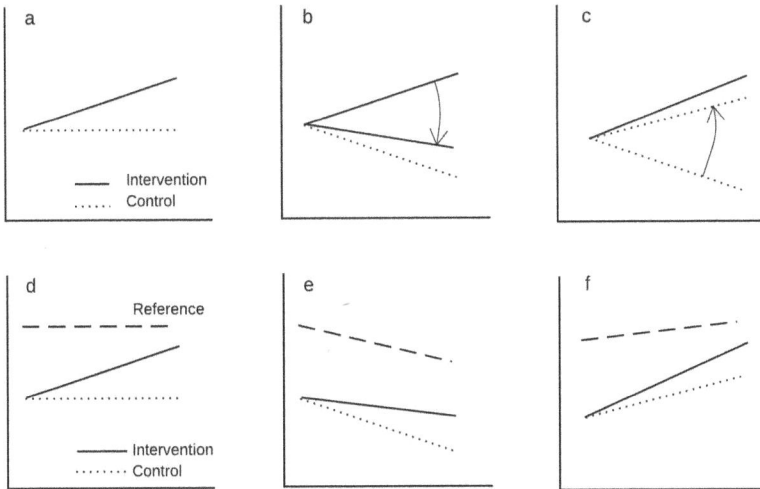

Figure 19.2: Simplified representation of mean agri-environment program outcomes, contrasting mean change over the course of investment in intervention sites (solid line) against mean background change (dotted lines, panels a–f), and mean change in reference sites (dashed lines, panels d–f).

Source: Authors' research.

For example, the most recent State of the Environment Committee Report (2011) concluded that native ecosystems on private land are mostly in decline. This suggests that agri-environment schemes could be considered successful even if managed sites show a reduced decline in condition compared with non-intervention sites (e.g. Figure 19.2b).

Another possibility is that positive changes in the extent of native ecosystems, consistent with the objectives of agri-environment schemes, are occurring spontaneously due to declining extent of agricultural production and land use changes (Kyle and Duncan 2012; Geddes et al. 2011). Government investment in agri-environment schemes may have only marginal benefit over the improving background trend (e.g. Figure 19.2c).

Counterfactual data from control sites — or at minimum a coherent conceptual model of the presumed fate of control sites (see the final option of Table 19.1) — are required to interpret responses measured at intervention sites and appropriately evaluate the success or otherwise of a given agri-environmental investment scheme. As Figure 19.2d–f illustrates, an *additional* contrast against reference sites (the desired state) can provide valuable additional insight into the relative performance of interventions against non-intervention (Downes et al. 2002; Coffman et al. 2014).

There are different types of controls and contrasts

In the context of agri-environment schemes, distinct options for control sites exist that permit subtle differences in the inference possible (Table 19.1). It is important to think about what these options mean, and the limits to their interpretation. Here we present a range of different control options available for comparison, and assess each against its likely relative value with regards (1) strength of inference for a given sample size, (2) transferability of the inference beyond the specific program, (3) contribution to a causal understanding, and (4) capacity to accommodate multiple response variables.

The most efficient learning scenario is to construct a management experiment where there is potential to select sites based on a management question, and then randomly allocate sites to treatment and non-treatment classes. This scenario, for a given sample size, offers the strongest chance to learn about what works, where, and why. However, it can be hard (and often impossible) to get support and sufficient sample sizes for experimentation on large scales, particularly involving public money on private land. In the past, many Australian states and the CSIRO had access to publicly owned production land where demonstration farming and experimentation took place for agricultural productivity. Such holdings may offer a cheaper and more secure opportunity for the Australian Government to learn about the effectiveness of interventions in comparison to building management experiments into the implementation of agri-environment schemes.

Table 19.1: Attributes and constraints of distinct counterfactual and contrast scenarios, spanning tailored management experiments where the strongest inference might be anticipated through to model-based calculation of impact, for which the major constraint is the paucity of available evidence from the types of sampling listed higher in the table.

		Attributes relating to inference				Constraints	
Type of control / contrast		Strength	Transferability	Causal understanding	Multiple responses	Convenience	Reliability
Management experiment		stronger				L	L
Field sampling controls	Paired site on winning bid site					H	H
	Unsuccessful bid					M	M
	Random sample					L	L
contrast	Space for time substitution					M	L
	Reference (contrast against desired state)					H	H
Model	Model based counterfactual	weaker				H	H

Source: Authors' research.

A more likely route to ensure adequate sample sizes is to build quantitative evaluation into major investment programs. In this instance, the researcher has to reactively match control sites to sites selected by the funding agency. Lindenmayer et al. (2012), for example, have established a major evaluation of the impact of 10–15-year grazing management agreements in the Box Gum Grassy Woodlands on native flora and fauna, comparing treatment sites with control sites located on the property of winning bidders in the auction program. This approach takes advantage of the convenience of an established relationship with the participating landholder, and should reduce sources of random variation, such as spatial variation in environmental factors, and farm level management factors (both historical and contemporary). These benefits permit a relatively robust comparison of the difference in change between treatments and sites under pre-existing management, although some limitations exist (see below).

Duncan and Vesk (2013) suggested that unsuccessful bidders for agri-investment payments could be a potential source of control sites for successful bids. These sites have the advantage of being assessed as part of the bidding process, so landholders have been engaged and some data may be available to guide their inclusion. However, greater random difference due to spatial effects and past management might be expected, so the comparison would be expected to be noisier, and the inference weaker, for a given sample size compared to a matched control on a winning bid. In both of the preceding cases, it would be hard to match the starting condition of funded sites with control sites, as the selection process is designed to favour the best-quality sites available for intervention.

The potential downside of locating control sites on the property of a paid participant is that the existence of the treatment, and the accompanying negotiation, may bias a participant's approach to management. After all, permanent behavioural and attitudinal change is one objective of government investment in agri-investment schemes, rather than merely switching on favourable management for the duration of a contract. Lindenmayer et al. (2012) expressed confidence the management of control sites was unaffected by the management of the paired treatment, but elsewhere involvement in auctions has been shown to influence the way landholders manage non-intervention areas (Windle et al. 2009).

These comparisons do not enable change in treatment sites to be compared against the way the average site is managed, but rather against the way a landholder positively disposed to conservation programs might manage their land. The estimates of background change we obtain from controls in agri-environment schemes are therefore unlikely to represent the average trend from the broader landscape, which may constitute a more desirable impact statement for program managers. We might ideally like to randomly sample appropriate controls for funded treatment units from the broader landscape, however, in practice problems of selection bias remain in those that choose to allow their properties to be visited and sites sampled. An estimate of variation associated with a randomly selected group of control sites would also require a larger sample size.

The minimal option should be an explicit, model-based counterfactual comparison (see final option of Table 19.1). A simple version might involve sampling intervention sites, with the amount of change being subsequently claimed as impact against a clear justification for what control sites are expected to do. The most important change from the way evaluation is typically conducted at present is that the model of control site behaviour must be made explicit and well justified. We are not aware of any examples of this kind of practice in action. More commonly, change data from intervention sites is claimed as an impact, with no declaration about what is presumed to be happening under business-as-usual scenarios.

One can imagine further sophistications of this model-based approach, where relevant quantitative data on background change under business-as-usual scenarios can be used to simulate the likely behaviour of control scenarios in evaluation. However, this possibility will not generally be available or particularly compelling until relevant data is accumulated from the more conventional approaches higher up in Table 19.1.

Why are controls and counterfactual contrasts so rarely obtained?

There are challenges to be overcome in identifying, negotiating, funding, and interpreting appropriate counterfactual data to use in evaluating agri-environmental schemes (e.g. Ferraro 2009; Kibler 2011). Perhaps due to these challenges, program specific or generic monitoring and evaluation advice relevant to agri-environment schemes may be released without any mention of their importance, or considerations for sampling controls or counterfactual data (e.g. the Conservation Measures Partnership 2013; DSEWPaC 2013). Such omissions contribute to evaluation design that does not explicitly discuss how counterfactual evidence will be gathered, inferred, or done without. We agree with Ferraro (2009) that evaluation plans should at least demonstrate counterfactual thinking.

Cost of sampling control sites

One factor that surely limits agency support for sampling control sites is cost, usually conceived as an increase in the total monitoring budget. A simplistic assumption might be that including (paired) controls in the sampling design would double the cost allocated to monitoring, further diminishing the amount assigned to action. However, one of the primary reasons to monitor is to demonstrate impact and learn about treatment effectiveness (e.g. this program has increased the occupancy of woodland birds by X per cent over a background trend of Y per cent). Therefore, it can readily be demonstrated that control sampling is fundamental to all cost-effective designs, as no amount of monitoring without controls can support the required inference.

Once the desired result statements are clearly defined, one can simulate the data collection and analysis in advance to examine which sampling scenarios are mostly likely to cost-effectively deliver that result. In reality, strategic sampling for impact evaluation, by including a subset of treatments and controls, could probably be achieved for a similar cost to that typically spent on monitoring programs which have failed to generate strong insight about the effectiveness of interventions.

Funding models and process

The way in which agencies tend to allocate and deliver funding for agri-environmental schemes, at both program and project level, can make it difficult to design and implement strong quantitative evaluation. For example, projects may be awarded funds concurrent with, or even before, the design of evaluation, making it impossible to obtain pre-intervention data from intervention sites, let alone control sites.

Lack of clarity of objectives and process model hinders evaluation design

The failure to make explicit program objectives and a conceptual or process model of cause and effect hinders monitoring of any sort (Field et al. 2007), and compounds the difficulty in identifying an appropriate control in program evaluation. The objectives and assumptions in the conceptual model of cause and effect should indicate what sort of trajectories and effect sizes to expect, and also guide the selection of a suitable control.

Some investment objectives are particularly difficult to control for. While controls should be achievable for site-scale interventions and responses (e.g. Lindenmayer et al. 2012; Hale et al. 2014), the greater the spatial scale of the program objective (e.g. landscape connectivity), or the greater the number of links in the causal chain between source and impact (e.g. changing agricultural land use to reduce oceanic hypoxia, Rabotyagov et al. 2014), the harder it may be to identify appropriate controls.

Horses for courses in counterfactual data collection

We intuitively think of the ideal control site as an independent site, closely matched to our treatment. Constructive solutions could be identified by focusing instead on the specific counterfactual requirements for response variables in the conceptual model of cause and effect. Within a given investment program, this may imply different data gathered at different scales or indeed locations. An appropriate control for measuring aquatic responses should differ from terrestrial plant responses, which could be different again for faunal responses.

Consider investment in restoration of riparian corridors. The impact of the investment on terrestrial vegetation might be well accommodated by a fenceline contrast (i.e. comparison of management regime either side of a fence), whereas the control for aquatic responses may be best placed some distance upstream of the treatment to maximise independence owing to the directional movement of water and its constituents. For occupancy responses of mobile biota, direction may not matter, but distance between sites may be important to achieve requisite independence.

What is required to improve our understanding?

The design of evaluation for all agri-environmental schemes should explicitly include counterfactual thinking (Ferraro 2009). In theory, this thinking should be represented in program logic diagrams or conceptual models that set out the expected difference in outcome comparing intervention and no-intervention (e.g. model-based counterfactual, Table 19.1). Larger schemes should produce well designed and resourced quantitative evaluation, linked to those models.

Be realistic. High rigour generally means less replication and reduced coverage of important contexts or covariates. A strategic mix of observational and experimental studies that explicitly complement and reference each other are required.

Large and small agri-environment schemes should do the best with what is available, including supporting post hoc comparisons, and using simulation models and scenario analyses. All techniques that can help make the most of existing data will remain important, and sound evaluation design should inform and guide data requirements. However, none of these fallback options excuse the persistent failure to conduct robust evaluation of agri-environment programs, including obtaining counterfactual data.

Be upfront about limitations to interpretation. Where less than ideal evaluation and assessment takes place, it is important to clearly state the limits to interpretation. For example, Duncan and Vesk (2013) estimated a substantial reduction in weed cover in sites funded by Victoria's BushTender program comparing before and after data from intervention sites. However, due to the lack of control sites, they explicitly cautioned that the observed changes were just as plausibly attributable to sustained drought.

Synthesise and disseminate. There are major programs beginning to establish rolling synopses of evidence of effects of different agri-environmental interventions (Dicks et al. 2013; Pullin and Knight 2009), including studies containing counterfactual evidence. Those synopses are tailored to the implementation context of northern and western Europe, so Australia should expect to support its own version, given our environmental, cultural and land use history and pattern.

Conclusions and recommendations

The current forms of monitoring and reporting (e.g. MERI — Monitoring, Evaluation Reporting, Improvement — Australian Government Land and Coasts 2009) undertaken in Australia have a valid role in the delivery and evaluation of agri-environmental schemes, but there is an urgent need to translate rhetoric into disciplined practice in quantifying environmental impact. However, our current systems routinely deliver poorly designed data collection activities, the results of which are scarcely, if ever, analysed and publicised.

Considerable coordination and nuance may be required to obtain inference about the impact of interventions in a cost-effective manner. For example, counterfactual data for interventions may be sourced at different spatial and temporal scales, as defined by the conceptual model relationship between treatment and response. It is likely that no evaluation program will encompass all elements and scales of space and time, but every program should be expected to make a coherent statement about effectiveness that includes an explicit contrast with a non-intervention scenario.

Done well, effective evaluation incorporating counterfactual data need not cost more than is currently expended on monitoring and evaluation. Importantly, even though not all programs will undertake such sampling, all should explicitly represent counterfactual thinking in MERI plans and program design. In addition to an immense literature relevant to setting objectives for agri-environment schemes, we offer the following checklist for evaluating whether a MERI plan for an agri-environment scheme has met minimum requirements:

1. The management behaviour or resource trend that funded treatments are intended to address, ameliorate, or reverse should be specified, in its appropriate spatial and temporal context.
2. The counterfactual prognosis (in terms of averages and some indication of variation) should be specified for the term of the funded treatments, and beyond, according to the definition of 1.
3. Elements 1 and 2 should be expressed in a manner that conveys the degree of certainty and scientific consensus, regarding averages and sources of variation, so that MERI programs that will guide field data collection are designed for maximum benefit.

Acknowledgements

We thank Steve Sinclair for discussions and suggestions that improved an earlier draft. Dean Ansell and anonymous reviewers provided ideas and suggestions that further improved our final manuscript.

References

Australian Government Land and Coasts (2009) *NRM MERI framework: Australian Government natural resource management monitoring, evaluation, reporting and improvement framework*, Department of the Environment, Water, Heritage and the Arts, Canberra.

Box, G. and G. Tiao (1975) 'Intervention analysis with applications to economic and environmental problems', *Journal of the American Statistical Association* 70(349): 70–9. Available at: www.tandfonline.com/doi/abs/10.1080/01621459.1975.10480264.

Coffman, J.M., et al. (2014) 'Restoration practices have positive effects on breeding bird species of concern in the Chihuahuan Desert', *Restoration Ecology* 22(3): 336–44. Available at: doi.wiley.com/10.1111/rec.12081.

The Conservation Measures Partnership (2013) *Open Standards for the Practice of Conservation*, version 3. Available at: www.conservationmeasures.org.

Department of Sustainability, Environment, Water, Population and Communities (DSEWPaC) (2013) *Biodiversity Fund: Ecological monitoring guide*, Commonwealth of Australia, Canberra. Available at: www.environment.gov.au/cleanenergyfuture/biodiversity-fund/meri/pubs/eco-monitoring-guide.pdf.

Dicks, L.V., et al. (2013) 'A transparent process for "evidence-informed" policy making', *Conservation Letters* 7(2): 119–25. Available at: doi.wiley.com/10.1111/conl.12046.

Downes, B.J., et al. (2002) *Monitoring ecological impacts: Concepts and practice in flowing waters*, Cambridge University Press, New York.

Duncan, D.H. and P. Vesk (2013) 'Examining change over time in habitat attributes using Bayesian reinterpretation of categorical assessments', *Ecological Applications* 23(6): 1277–87. Available at: www.esajournals.org/doi/abs/10.1890/12-1670.1.

Ferraro, P.J. (2009) 'Counterfactual thinking and impact evaluation in environmental policy', *New Directions for Evaluation* (122): 75–84.

Ferraro, P.J. and S.K. Pattanayak (2006) 'Money for nothing?: A call for empirical evaluation of biodiversity conservation investments', *PLoS Biology* 4(4): e105.

Field, S.A., et al. (2007) 'Making monitoring meaningful', *Austral Ecology* 32(5): 485–91. Available at: doi.wiley.com/10.1111/j.1442-9993.2007.01715.x.

Geddes, L.S., et al. (2011) 'Old field colonization by native trees and shrubs following land use change: Could this be Victoria's largest example of landscape recovery?' *Ecological Management and Restoration* 12(1): 31–6. Available at: doi.wiley.com/10.1111/j.1442-8903.2011.00570.x.

Hajkowicz, S. (2009) 'The evolution of Australia's natural resource management programs: Towards improved targeting and evaluation of investments', *Land Use Policy* 26(2): 471–8. Available at: dx.doi.org/10.1016/j.landusepol.2008.06.004.

Hale, R., et al. (2011) *Assessing ecological indicators for monitoring responses to riparian restoration in lowland streams of the southern Murray-Darling Basin*, Murray-Darling Basin Authority Project Report MD606, Monash University and Arthur Rylah Institute for Environmental Research.

Hale, R., et al. (2014) 'Bird responses to riparian management along degraded lowland streams', *Ecological Restoration* 23(2): 104–12. DOI:10.1111/rec.12158.

Kibler, K.M., D.D. Tullos and G.M. Kondolf (2011) 'Learning from dam removal monitoring: Challenges to selecting experimental design and establishing significance of outcomes', *River Research and Applications* 27: 967–75.

Kyle, G. and D.H. Duncan (2012) 'Arresting the rate of land clearing: Change in woody native vegetation cover in a changing agricultural landscape', *Landscape and Urban Planning* 106(2): 165–73. Available at: linkinghub.elsevier.com/retrieve/pii/S0169204612000916.

Lindenmayer, D.B., et al. (2012) 'A novel and cost-effective monitoring approach for outcomes in an Australian biodiversity conservation incentive program', *PLoS ONE* 7(12): e50872. Available at: www.plosone.org/article/info:doi/10.1371/journal.pone.0050872#s1.

Margoluis, R., et al. (2009) 'Design alternatives for evaluating the impact of conservation projects', *New Directions for Evaluation* 2009(122): 85–96. Available at: onlinelibrary.wiley.com/doi/10.1002/ev.298/abstract.

Mascia, M.B., et al. (2014) 'Commonalities and complementarities among approaches to conservation monitoring and evaluation', *Biological Conservation* 169: 258–67. Available at: linkinghub. elsevier.com/retrieve/pii/S0006320713003960.

Petticrew, M. and H. Roberts (2003) 'Evidence, hierarchies, and typologies: Horses for courses', *Journal of Epidemiology and Community Health* 57(7): 527–9. Available at: www.pubmedcentral. nih.gov/articlerender.fcgi?artid=1732497&tool=pmcentrez&rend ertype=abstract.

Prober, S., R. Standish and G. Wiehl (2011) 'After the fence: Vegetation and topsoil condition in grazed, fenced and benchmark eucalypt woodlands of fragmented agricultural landscapes', *Australian Journal of Botany* 59(4): 369–81. Available at: www.publish.csiro. au/?paper=BT11026.

Pullin, A.S. and T.M. Knight (2009) 'Doing more good than harm: Building an evidence-base for conservation and environmental management', *Biological Conservation* 142(5): 931–4. Available at: www.sciencedirect.com/science/article/pii/S0006320709000421.

Rabotyagov, S., et al. (2014) 'Robust optimization of agricultural conservation investments to cost-efficiently reduce the northern Gulf of Mexico hypoxic zone', *Proceedings of the World Congress of Environmental and Resource Economists*, 28 June – 2 July, Istanbul, Turkey, pp. 1–36.

Read, C.F., et al. (2011) 'Surprisingly fast recovery of biological soil crusts following livestock removal in southern Australia', *Journal of Vegetation Science* 22(5): 905–16. Available at: doi.wiley. com/10.1111/j.1654-1103.2011.01296.x.

Sarkar, S., et al. (2006) 'Biodiversity conservation planning tools: Present status and challenges for the future', *Annual Review of Environment and Resources* 31(1): 123–59. Available at: www.annualreviews.org/ doi/abs/10.1146/annurev.energy.31.042606.085844.

State of the Environment 2011 Committee (2011) *Australia state of the environment 2011*, independent report to the Australian Government Minister for Sustainability, Environment, Water, Population and Communities, Canberra.

Stem, C., et al. (2005) 'Monitoring and evaluation in conservation: A review of trends and approaches', *Conservation Biology* 19(2): 295–309. Available at: onlinelibrary.wiley.com/doi/10.1111/j.1523-1739.2005.00594.x/full.

Stewart-Oaten, A.J. and Bence (2001) 'Temporal and spatial variation in environmental impact assessment', *Ecological Monographs* 71(2): 305–39. Available at: www.esajournals.org/doi/pdf/10.1890/0012-9615(2001)071[0305:TASVIE]2.0.CO;2.

Stoffels, R. and K. Weatherman (2014) *The decommissioning of Lake Mokoan: Effects on water quality and fishes of the Broken River*, final report prepared for the Goulburn Broken Catchment Management Authority, Wodonga.

Thomson, J.R.J., et al. (2010) 'Bayesian change point analysis of abundance trends for pelagic fishes in the upper San Francisco Estuary', *Ecological Applications* 20(5): 1431–48. Available at: www.esajournals.org/doi/abs/10.1890/09-0998.1.

Windle, J., et al. (2009) 'A conservation auction for landscape linkage in the southern Desert Uplands, Queensland', *The Rangeland Journal* 31(1): 127. Available at: www.publish.csiro.au/?paper=RJ08042.

20

Achieving greater gains in biodiversity from agri-environment schemes

Philip Gibbons

Key lessons

- Agri-environment schemes should focus on investments that maximise gains in biodiversity relative to the status quo.

- It is widely viewed that our conservation priorities should be biodiversity that has high values of irreplaceability (biodiversity that must be protected to achieve conservation targets) and has high vulnerability (the likelihood that biodiversity will be lost without conservation investment).

- Too much agri-environmental investment is in biodiversity that is not vulnerable to loss, and this investment does not result in substantial gains to biodiversity relative to the status quo.

- Agri-environment schemes should be informed by a decision framework that calculates gains in biodiversity relative to the status quo (i.e. the difference in biodiversity with investment and losses without investment). This will shift more investment from large, high-quality remnants to smaller, more modified remnants that are more vulnerable to loss.

Figure 20.1: Increasing the conservation status of threatened species such as the hooded robin (*Melanodryas cucullata*) is a key motivation for many agri-environment schemes.
Source: Photo by Geoff Park.

If I was to invest in a farm, I would not invest in the largest farm on the market, but a farm with greatest potential to return growth on my investment. In this chapter, I argue that our current investment strategy for biodiversity in agri-environment schemes is frequently the opposite: we tend to invest in sites that support large amounts of biodiversity, rather than sites with potential to return the greatest gains in biodiversity as a result of the investment (Figure 20.2).

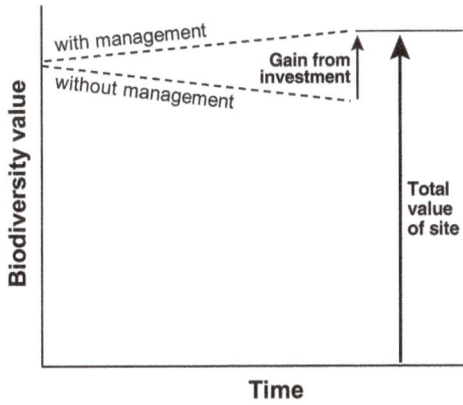

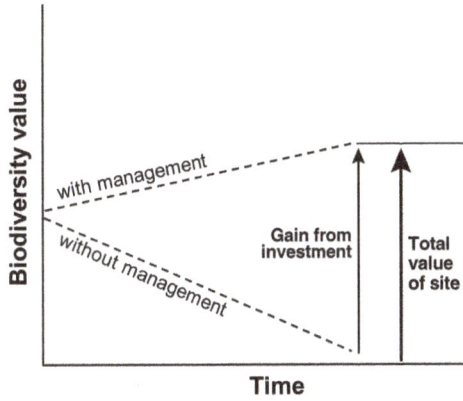

Figure 20.2: Agri-environment schemes often prioritise sites for investment based on the total biodiversity that occurs on a site, rather than the gain in biodiversity that is likely to occur with investment. The former approach places priority for investment in high-quality sites (top), while the latter places priority on moderate-quality sites (bottom).

Source: Author's research.

The theory of biodiversity prioritisation

Historically, there has been a bias in protected areas within Australia towards land unsuitable for agriculture. Australia's reserves therefore tend to protect areas that are steep and/or infertile (Pressey et al. 2002). As a consequence, one-third of Australia's bioregions are very poorly represented (less than 5 per cent) in Australia's National Reserve

System (Hatton et al. 2011). Many of the most poorly protected ecosystems in Australia occur in agricultural landscapes. Margules and Pressey (2000) outline the principles upon which future conservation investment should be based to rectify past biases. These and other authors argue that our conservation priorities should be those areas, or elements of biodiversity, that have highest values of irreplaceability and vulnerability (Figure 20.3).

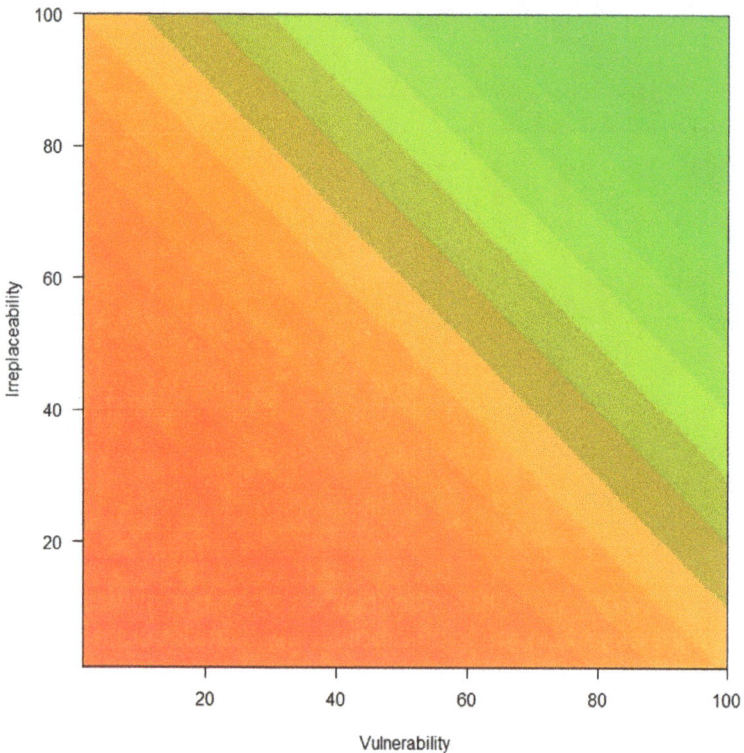

Figure 20.3: Higher priorities for conservation are those sites that have high irreplaceability and high vulnerability, as indicated in green.
Source: Author's research.

Irreplaceability refers to the likelihood that a site must be protected to meet a conservation target. This concept is analogous to complementarity (Margules and Pressey 2000), which is the extent to which an area contributes unrepresented biota to a reserve system. For example, if the conservation target is to protect at least 17 per cent of the original (pre-European) extent of each ecosystem, and less than

17 per cent remains of a particular ecosystem, then all remaining remnants of this particular ecosystem are 100 per cent irreplaceable. That is, all examples of this ecosystem that are left are required to meet the conservation target.

Vulnerability refers to the likelihood that a site will be lost without immediate conservation investment (Margules and Pressey 2000). This principle is based on the assumption that the areas and actions needed to secure biodiversity exceed the resources available for conservation and thus we need to prioritise our conservation investment in those areas or for those actions that are most urgent. For example, a remnant in which natural regeneration has ceased due to a land use such as livestock grazing — and will disappear over time — is more vulnerable than a remnant in which natural regeneration is still occurring.

Where is our current biodiversity investment in agri-environment schemes?

The principle of irreplaceability has been embraced within Australia. We now tend to prioritise our conservation investment in biota that are poorly represented in our existing protected area network (National Reserve System Task Group 2009) and this thinking is reflected in agri-environment schemes. For example, the Environmental Stewardship Program (see Chapter 3) focused on ecological communities poorly represented in Australia's National Reserve System. However, the concept of vulnerability has not been embraced as fully. For example, BushTender is the most established agri-environment scheme focused on biodiversity in Australia, with over 35,000 ha managed for conservation on private land. The metric that underpins the way sites are prioritised for investment in BushTender multiplies the current condition of a site by the condition that the site is likely to achieve with investment (Maron et al. 2013). That is, other things being equal, a site that can be improved from 20 to 80 out of 100, will score less ($20 \times 80 = 1,600$) than a site that can be improved from 70 to 90 out of 100 ($70 \times 90 = 6,300$) even though the gain in biodiversity at the latter site (20) is a third of the gain achieved on the former site (60). In this metric, the total value of the site for biodiversity is given more weight than the gain in biodiversity that will be achieved. If the metric was instead based only on the gain in biodiversity at each site, then a site that can be improved from 20 to 80 of 100 (with a gain of 60) will be prioritised for investment over a site that can be improved from 70 to 90 (with a gain of 20). These two scenarios are illustrated conceptually in Figure 20.2. Similarly, the investment strategy underpinning the Australian Government's Environmental Stewardship Program for Box Gum Grassy Woodland in south eastern Australia stipulated that remnants must be at least 10 ha. Focusing agri-environmental investment on patches greater than 10 ha could result in most of the remaining area of this ecological community being lost. This is because most of this ecological community (at least in the southern part of its range) occurs in much smaller patches (Gibbons and Boak 2002), and these smaller patches are more likely to be approved for clearing (Gibbons et al. 2009) and are least likely to contain natural regeneration (Weinberg et al. 2011). Maron et al. (2013) identifies other metrics that result in similar outcomes. Given this, our agri-environmental investment may not be returning the gains in biodiversity that are possible.

Where should we be investing?

There is a bias in agri-environmental investment towards large, high-quality remnants for three key reasons: (1) there is an assumption that all remnants on private land are equally vulnerable to loss; (2) there is an assumption that small remnants and poor-quality remnants are of limited value for conservation relative to large remnants; and (3) managing fewer, larger remnants is considered more cost-effective than managing more, small remnants. These assumptions should be considered critically.

On the first point — the assumption that all remnants on private land are equally vulnerable to loss — not all biodiversity on private land that is highly irreplaceable is also vulnerable to loss. For example, the larger remnants of our most cleared ecosystems have not been cleared because they tend to occur on sites with low productivity and are therefore neither intensively grazed, nor profitable to clear or fertilise, and are not as vulnerable to invasion by exotic plants than more productive sites. It is also important to acknowledge that there is legislation that affords a high level of protection to these sites, and it is important that agri-environment schemes do not undermine this existing duty of care. In contrast, smaller remnants of these ecological communities are very vulnerable to loss because they tend to be on more productive land, under greater threat from land management practices in the agricultural matrix (Driscoll et al. 2013), have a lower likelihood of supporting natural regeneration (Weinberg et al. 2011), and are more likely to be approved for clearing (Gibbons et al. 2009).

On the second point, there is a widely held view that small sites, poor-quality sites, and highly fragmented landscapes are of limited value for conservation. For example, the Habitat Hectares metric (Parkes et al. 2003) that underpins agri-environmental investment in Victoria affords greater score or value to larger patches and more intact landscapes. However, the species-area curve (Rosenzweig 1995) predicts that adding a unit area of habitat to a smaller patch or more fragmented landscape will result in a greater increase in species than improving or adding the same amount of habitat to a larger patch or intact landscape (Figure 20.4). This prediction has been recently confirmed for birds in agricultural landscapes within Australia (e.g. Cunningham et al. 2014; Huth and Possingham 2011). A counter argument is that conservation efforts in small remnants will favour only common species. While this argument might hold for dispersal-limited species, small patches make an important contribution to the conservation of declining birds (Fischer and Lindenmayer 2002).

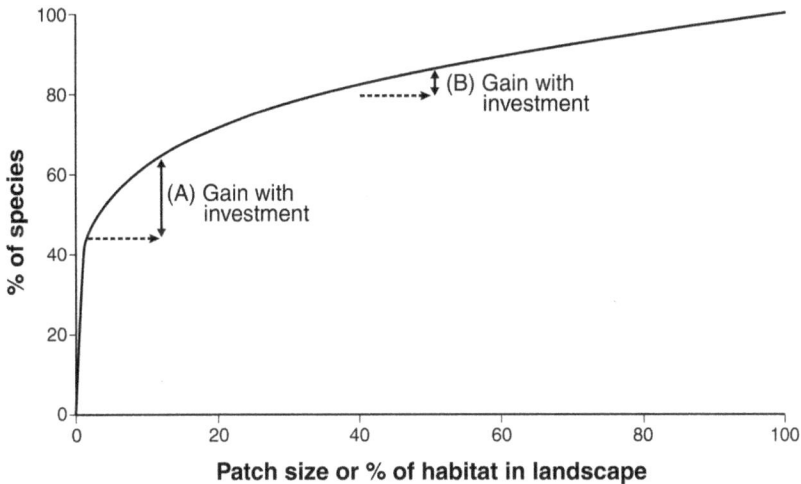

Figure 20.4: A generic species-area curve (using an exponent of 0.25) indicating the predicted gain in species richness with the same conservation effort (indicated by arrows) in (A) a smaller patch or highly cleared landscape compared with (B) a larger patch or more intact landscape.

Source: Author's research.

Finally, large sites are considered to be more cost-effective to manage than small sites. However, if the cost-effectiveness of investments are considered in agri-environment schemes (i.e. biodiversity gain is divided by cost), then a higher cost of management (per ha) on smaller sites could be offset by a higher rate of gain in biodiversity. Results from the Australian Government's Environmental Stewardship Program found that the cost (per ha) of tenders to manage biodiversity on private land were similar for small and large remnants (Whitten et al. 2009). That is, no discernible economy of scale was observed. This may be because the unit of management for agri-environment schemes is effectively the farm rather than the individual remnant — and thus is influenced by characteristics of the farming enterprise rather than the individual remnant. However, more sites may be more costly to administer (e.g. more assessments) or manage where infrastructure (e.g. fencing) is required.

Conclusion

Agricultural areas of Australia support many highly irreplaceable ecosystems that are vulnerable to loss, and these landscapes are important for biodiversity conservation. Investment in biodiversity conservation through agri-environment schemes makes intuitive sense. However, the amount of investment in agri-environment schemes is unlikely to be sufficient to meet our conservation targets, so we must prioritise this investment carefully. Shifting the focus of agri-environmental investment towards activities that are likely to return the greatest gains in biodiversity (i.e. the difference with investment and losses without investment) is likely to see a greater emphasis on investment in smaller, modified remnants and more fragmented landscapes than is currently the case.

References

Cunningham, R.B., D.B. Lindenmayer, M. Crane, et al. (2014) 'The law of diminishing returns: Woodland birds respond to native vegetation cover at multiple spatial scales and over time' *Diversity and Distributions* 20: 59–71.

Driscoll, D.A., S.C. Banks, P.S. Barton, D.B. Lindenmayer and A.L. Smith (2013) 'Conceptual domain of the matrix in fragmented landscapes', *Trends in Ecology and Evolution* 28: 605–13.

Fischer, J. and D.B. Lindenmayer (2002) 'Small patches can be valuable for biodiversity conservation: Two case studies on birds in southeastern Australia', *Biological Conservation* 106: 129–36.

Gibbons, P. and M. Boak (2002) 'The value of paddock trees for regional conservation in an agricultural landscape', *Ecological Management and Restoration* 3: 205–10.

Gibbons, P., S. Briggs, D. Ayers, et al. (2009) 'An operational method to assess impacts of land clearing on terrestrial biodiversity', *Ecological Indicators* 9: 26–40.

Hatton, T., S. Cork, P. Harper, et al. (2011) 'Australia state of the environment 2011', independent report to the Australian Government Minister for the Environment and Heritage, Canberra.

Huth, N. and H.P. Possingham (2011) 'Basic ecological theory can inform habitat restoration for woodland birds', *Journal of Applied Ecology* 48: 293–300.

Margules, C.R. and R.L. Pressey (2000) 'Systematic conservation planning', *Nature* 415: 243–53.

Maron, M., J.R. Rhodes and P. Gibbons (2013) 'Calculating the benefit of conservation actions', *Conservation Letters* 6: 359–67.

National Reserve System Task Group (2009) *Strategy for Australia's National Reserve System 2009–2030*, Department of the Environment, Water, Heritage and the Arts, Canberra.

Parkes, D., G. Newell and D. Cheal (2003) 'Assessing the quality of native vegetation: The "habitat hectares" approach', *Ecological Management and Restoration* 4: S29–S38.

Pressey, R., G. Whish, T. Barrett and M. Watts (2002) 'Effectiveness of protected areas in north-eastern New South Wales: Recent trends in six measures', *Biological Conservation* 106: 57–69.

Rosenzweig, M.L. (1995) *Species diversity in space and time*, Cambridge University Press, Melbourne.

Weinberg, A., P. Gibbon, S.V. Briggs and S.P. Bonser (2011) 'The extent and pattern of *Eucalyptus* regeneration in an agricultural landscape', *Biological Conservation* 144: 227–33.

Whitten, S.M., R. Gorddard, A. Langston and A. Reeson (2009) *A review of the Box Gum Grassy Woodlands Environmental Stewardship Project Metric*, report for the Australian Government Department of the Environment, Water, Heritage and the Arts, CSIRO Sustainable Ecosystems, Canberra.

21

Lessons for policy from Australia's experience with conservation tenders

Graeme Doole and Louise Blackmore

Key lessons

- This chapter reviews the drivers of cost-effectiveness in conservation tender programs, with a focus on what promotes landholder participation, based on survey results.

- Key lessons are drawn from the statistical analysis of survey responses from landholders, researchers, and agency staff with previous involvement in these schemes.

- Non-landholders identify the value of flexible tender designs, involving low-cost monitoring and strong relationships with stakeholders.

- Landholder responses suggest that tender schemes may have limited impact due to the crowding out of private investment, inadequate support during bidding and monitoring, and high administrative workload.

- Overall, future tender programs must employ options to counteract crowding out if they are to achieve additional environmental outcomes.

Australia is one of 17 mega-diverse nations, which overall support more than 70 per cent of the world's biodiversity, but constitute less than 10 per cent of the global land area (Chapman 2009). The existing suite of protected areas found throughout Australia is inadequate to achieve broad biodiversity conservation, especially in regions highly favoured for agricultural production (Fitzsimmons and Wescott 2001). Over the past decade, there has been an increasing focus on conserving national biodiversity, broadly promoted in recent years through Australia's Biodiversity Conservation Strategy 2010–2030.

In particular, a strong motivation to achieve more cost-effective environmental outcomes on private land has stimulated the widescale use of market-based instruments (MBIs). Conservation tender (CT) schemes have been the most widely applied MBIs for biodiversity conservation throughout Australia to date. CT programs involve landholders submitting a bid to a management authority (e.g. regional natural management bodies or non-government organisations) outlining the actions they will take for a given payment. The authority evaluates the environmental benefit accruing to each set of actions and funds those landholder actions that will provide the greatest environmental benefit for the given budget (Latacz-Lohmann and Schilizzi 2005; Windle and Rolfe 2008).

The primary objective of this chapter is to review what drives the cost-effectiveness of CT programs, particularly relating to promoting the participation of landholders. Given the scarcity of available funds, it is justified that regulatory authorities seek to maximise the cost-effectiveness of programs to motivate biodiversity conservation among Australian landholders (Pannell 2008). It is also consistent with the broad goal of Australia's Biodiversity Conservation Strategy 2010–2030 to deliver conservation initiatives in a cost-effective manner. A cost-effective or efficient program is defined as one that provides a given set of conservation outcomes at least cost. It is timely to reflect on the drivers of the cost-effectiveness of these programs, given that despite the broad employment of conservation auctions over the last decade throughout Australia, their use has now declined markedly. A number of potential reasons for this disadoption exist, including but not limited to thin markets, high administration costs for landholders and governing agencies, lack of political will to continue funding, and uncertainty for producers regarding the implications of their contracts. This study focuses on these drivers, among others, to explore what determines the cost-effectiveness of conservation auctions across Australia. This analysis is particularly timely, given that it reflects on past programs that arose during a period of proliferation for conservation auctions in Australia, using this information to guide how to gain the best return for public funds in future programs. These future programs are likely to be targeted, reinforcing the need for careful design. This summary provides a useful assessment of cost-effective design for auction programs, drawn from the Australian experience, for readers in other nations. Lessons here are drawn from the statistical analysis of survey responses from non-landholders (n=49) and landholders (n=23) who have been involved in past CT programs. More detail regarding these studies are available in Doole et al. (2014) and Blackmore and Doole (2013) respectively. Non-landholders are researchers and agency staff who have previously been involved in CT schemes, while the landholders are previous CT participants from north central Victoria. Low sample sizes motivate the use of small-sample econometric methods (generalised maximum entropy regression) to analyse the primary drivers of cost-effectiveness and participation in the non-landholder and landholder data sets respectively (Golan et al. 1996).

Results and discussion

Statistical analysis of the results from the non-landholder survey (Doole et al. 2014) highlights five key lessons for improving the cost-effectiveness of CT programs. Listed in decreasing order of importance, these are:

1. *Increase funding for both individual tenders and groups of tenders.* Respondents generally believed that the cost-effectiveness associated with tender implementation increased with their scale. This highlights the presence of increasing returns to scale, whereby the costs of tender implementation are believed to decline as tenders are more widely used, due to improved administrative efficiency and 'learning by doing' by the agency (Schilizzi and Latacz-Lohmann 2012). As experience and familiarity builds within agencies, it is thought that transaction costs will greatly decrease, allowing their cost-effectiveness advantages over flat rate subsidy schemes to be fully realised (Windle and Rolfe 2008). However, recent experience in Australia suggests that the development of this experience may

be severely challenged, primarily by a lack of political will and the funding opportunities required for the continued implementation of these policy mechanisms. Broader adoption will also lead to more widespread achievement of biodiversity outcomes, provided these programs are effectively managed. Nevertheless, the overall scale of a program depends on the number of contracts formed and the scale of individual bids. A higher number of contracts will impose additional variable costs, such as those associated with monitoring. However, through learning from experience, cost-effectiveness associated with the fixed costs of a program, and some variable costs — for example, the development of more efficient monitoring programs —may fall across time.

2. *Develop flexible tender designs for broad implementation.* Substantial effort has now been invested in the assessment of tender programs involving different components — for example, through testing the relative implications of the number of rounds involved, whether tenders are budget- or target-constrained, whether bids are open or sealed, whether bidder information is symmetric or asymmetric, and so on (e.g. Schilizzi and Latacz-Lohmann 2007; Reeson et al. 2011; Boxall et al. 2012). However, the broad implementation of more generic tender designs may promote ease of organisation, and thereby the cost-effectiveness of these programs, as opposed to the development of auction designs that may suit a given set of circumstances more specifically, but are costly to identify and more difficult to apply. This principle has been observed in practice, with the success of the BushTender program (Stoneham et al. 2003) being replicated throughout Australia, in programs such as EcoTender and RiverTender. Nevertheless, it is important to note that such examples are very localised, with a high concentration in Victoria, and the number of examples across Australia is now in steady decline. The reasons for this decline are manifold, but mainly revolve around funding challenges and political climate. For these reasons, it is important to recognise that while the down-scaling of auction programs will hamper learning by doing, the isolation of principles for cost-effective design identified throughout this chapter remain of key importance. Indeed, their significance is promoted when it is considered that a need for efficient delivery is even more pronounced when funds are scarce.

3. *Encourage greater landholder competition.* This result is consistent with strong theoretical (Latacz-Lohmann and van der Hamsvoort 1997) and empirical (Stoneham et al. 2003; Connor et al. 2008)

evidence that this is a key principle underlying the success of CT programs, relative to subsidy schemes. Indeed, a central premise of neoclassical economics is that greater competition within markets will encourage greater efficiency through dissolution of market power. Nevertheless, a CT represents an artificial market, in which a management authority is the single buyer of products put forward by landholders. Accordingly, greater participation is not an end in itself. Indeed, there is an optimal number of participants that depends on balancing the impact of participation on competitive behaviour, while also considering its concomitant effect on evaluation, monitoring, and other administrative costs, especially those imposed prior to contract selection (Whitten et al. 2013).

4. *Invest in means to identify low-cost monitoring strategies.* Respondents highlighted the critical importance of addressing the difficulty associated with measuring the benefits of tenders for conservation activity and/or outcomes. Value for money in conservation programs depends on whether actions are achieving on-ground change (Connor et al. 2008). Accordingly, the development of appropriate low-cost management strategies is of vital importance. This result has also been identified for European MBI programs (Latacz-Lohmann and Schilizzi 2005).

5. *Establish strong landholder relationships.* New programs for environmental protection are difficult to understand for many landholders because of their complexity (Pannell et al. 2006). Accordingly, the overall cost-effectiveness of tenders was believed to respond to investment by regulatory authorities in landholder education and relationships. However, investment in relationships with landholders was determined to be around three times more important than investment in education and support in the non-landholder survey. This reflects the central importance of landholders as potential suppliers of improved environmental outcomes within tender programs. It also represents a critical challenge to auction programs, in that while these relationships are critical to success, they are also costly in terms of time and money. However, there is anecdotal evidence that these costs are mostly imposed in the early stages of a program, when familiarity and the competency of landholders is lower. Accordingly, these represent an explicit example of costs that decrease with learning by doing.

Analysis of the results from the landholder survey using descriptive statistics (Blackmore and Doole 2013) highlights five key lessons for increased landholder participation within tender programs. Listed in decreasing order of importance, these are:

1. *Crowding out of private conservation investment may be significant in Australian CT programs.* Crowding out describes a situation where government funding drives down private funding, providing little or no additional benefit in terms of practical outcomes. In the context of conservation auctions, crowding out of private conservation investment infers that public funds provided for conservation expenditure through the tender mechanism merely offset private funding, providing little or no additional conservation benefit. There is only mild agreement among respondents that the time they had spent on conservation had increased during and after their participation in CT programs. Most respondents were conservation-minded, with 83 per cent of the sample stating that they were active members of at least two environmental groups. These results suggest that pre-existing levels of voluntary conservation effort among the sample were likely to be high. This is consistent with the findings of DSE (2006) and Moon et al. (2012), and suggests that crowding out may limit the cost-effectiveness of tenders throughout Australia. Efforts to engage landholders outside existing conservation communities may improve the conservation outcomes achieved by CT programs.

2. *Landholders are likely to be receptive to 10-year contracts within tender programs.* Shorter contracts (e.g. five years or less) often fail to achieve desired outcomes, whereas longer contracts (e.g. 15 years or more) represent a commitment beyond what landholders are willing to accept. This suggests a tension between landholders' desire to implement long-term conservation management strategies, and concerns relating to the limited tenure of farmers and the potential restriction of the options available to future generations farming that land.

3. *Landholders feel that agency performance has been adequate in the areas of providing introductory materials and advertising, specifying management actions, and payment scheduling.* These may be considered low-priority targets for future improvement.

4. *Landholders feel agency performance needs improvement in the areas of providing information workshops, providing support during bid*

construction, and providing information regarding bid selection.
Tender programs typically limit the quantity of information
provided to participants to minimise the possibility of collusion,
which involves landholders conspiring to submit inflated bids
(Latacz-Lohmann and Schilizzi 2005). In this study, being unable
to communicate with other participants, having limited support
in constructing bids, and being confused about the bid-selection
process left many landholders feeling isolated, frustrated, and
even betrayed by the agency. There is evidence that landholder
participants in the Environmental Stewardship Program have
experienced similar issues (MJA 2010; Zammit 2013). Our results
support those of Moon et al. (2012) and Whitten et al. (2007),
indicating a need to ensure that landholders receive assistance
and support to maintain or increase program participation, while
seeking to minimise collusion among them.

5. *Landholders feel that monitoring within tender programs requires
   improvement, with ongoing support during this phase an important
   step in better delivering management outcomes.* Most landholders
   submitted annual reports to the agency, and monitoring site visits
   were generally infrequent or non-existent. A lack of monitoring
   visits made most landholders feel isolated and unsure of their
   ability to complete contracted tasks. Low monitoring activity by
   agency employees reflects cost concerns and a wish to promote
   ownership of the scheme by landholders. However, our results
   support those of Morrison et al. (2008), suggesting that more active,
   frequent monitoring by agencies could enhance participation rates,
   while also helping to improve the likelihood that meaningful
   environmental outcomes are attained among participants.

It is important to discuss the amount of information provided
by regulatory authorities (see fourth point in the preceding list)
further, since it has significant implications for auction performance.
The distinction between information relating to the quality of
alternative bids (in terms of their ability to deliver conservation
outcomes) and that relating to price is critical. Information regarding
the suitability of proposed actions put forward by participants is
necessary for sensible bids, with both parties (sellers and buyers)
within a conservation auction program benefiting from this data.

In contrast, a lack of price information promotes competition among landholders; hence, management authorities responsible for the implementation of these schemes generally perceive it to be optimal to not provide such data. However, this action can reduce the probability that suitable practices are put forward for funding, while increasing the uncertainty faced by landholders regarding the construction of reasonable cost estimates. The latter can increase inefficiency above that experienced in subsidy programs, given that increasing bids will generally accompany greater uncertainty on the behalf of landholders (Latacz-Lohmann and van der Hamsvoort 1997).

The revelation of too much information can increase transactions costs and promote strategic bidding by landholders, which reduces competition and cost-effectiveness. Moreover, the provision of information workshops can strengthen landholder networks and potentially promote collusion. Survey results indicated that the agencies involved in implementing the programs that landholders had participated in generally provided less information, rather than more. Together with the inefficiency highlighted with a lack of information above, survey output suggests that these actions are actually hampering future participation in tender programs, given that one of the primary drivers of participation identified in the landholder survey is honest and respectful communication with the implementing agency (see below).

Analysis of the results from the landholder survey using small-sample econometric methods (Blackmore and Doole 2013) highlights five key lessons for increased landholder participation within tender programs. Listed in decreasing order of importance, these are:

1. *Landholders are more likely to participate in future CT programs when they have a strong, respectful, and continuous relationship with the implementing agency, involving face-to-face contact.* CT programs are more likely to succeed in areas characterised by strong, trusting agency–landholder relationships (Whitten et al. 2013). This is also thought to apply to the adoption of innovations more broadly (Pannell et al. 2006). Where such relationships are absent, personal engagement with landholders should be a high priority to ensure program success.

2. *Landholders are more likely to participate in future CT programs when the administrative load associated with participation is low.*

Streamlining the participation process is likely to enhance participation in future programs. This aspect is outlined in more detail in Chapter 16.

3. *Landholders with an altruistic attitude and strong conservation focus, with a relatively low focus on monetary outcomes, are more likely to participate in future CT programs.* Survey results indicate that 75 per cent of respondents would participate in future programs, but only 25 per cent of respondents believe that conservation tenders increase short-term income. These findings emphasise the significant potential for CT programs to crowd out voluntary conservation spending on Australian farms. Investigating strategies to engage landholders without these attributes will be critical to the achievement of broad-scale biodiversity outcomes. Accordingly, this has been a focus of the recent Environmental Stewardship Program (MJA 2010).

4. *Landholders are less likely to participate in future CT programs when they have the necessary skills to undertake conservation works independently.* The majority of respondents to the landholder survey highlighted that their primary driver to be involved in a CT program was not monetary, but rather altruistic and associated with a strong conservation ethic. Survey results indicate that landholders may become confident to manage an area for biodiversity conservation, independent of agency involvement, after their contracts expire. Ensuring that landholders leave programs with a high skill level and a realistic ongoing management plan could liberate funding to recruit new landholders into future CT schemes.

5. *Landholders are less likely to participate in future CT programs if they have received strong support and education in previous schemes.* Respondents generally had a strong conservation ethic and did not alter their conservation management efforts significantly during or after the CT program. Forty per cent of the sample found the cost of additional administrative work prohibitive. Accordingly, good support and education by the implementing agency was generally felt by participants to adequately equip them with the necessary skills to undertake future conservation projects, independent of agency involvement and the administrative burden associated with participation.

Conclusions

Conservation tenders have been broadly utilised over the last decade for biodiversity conservation throughout Australia. It is timely to review what drives the cost-effectiveness of these programs, particularly relating to promoting the participation of landholders, to inform the future implementation of biodiversity markets.

Non-landholders generally believed that the cost-effectiveness of tender programs could be encouraged through the broad implementation of flexible tender designs that involved low-cost monitoring strategies and strong relationships with landholders. However, landholders highlighted that the cost-effectiveness of these schemes may be hampered due to the crowding out of private conservation investment, inadequate support during the bidding process and monitoring stage, and the workload associated with administration. Survey results showed that landholders within the sample are mainly driven by a strong conservation ethic and altruism, rather than a focus on monetary outcomes. Accordingly, many of the landholders who were interviewed changed their conservation activity little due to their involvement in a tender program. This highlights that the current use of tenders may actually achieve little additional benefit in terms of concrete biodiversity gains.

It is imperative to address the potential crowding out of public funding in CT schemes if they are to achieve prospective efficiency gains. Several means could be used to do so. First, market-segmentation analysis (Morrison et al. 2011) could be used to identify the characteristics of a given landholder population. This data can then be used to inform how a tender could be structured to reduce the potential for crowding out. Second, monitoring is important to ensure that trades are based on activity that is currently occurring. Third, encouraging landholders to submit the proportion of the total cost of an activity that they expect the agency to cover and the proportion they expect to cover encourages the idea that cost-sharing is expected, even if such information is not used during bid selection (Windle and Rolfe 2007). Last, a reserve price should be set, so that over-priced bids — potentially inflated due to the substitution of private investment by agency funds — are not accepted.

Acknowledgements

The authors would like to acknowledge funding and support received from National Environmental Research Program and the Centre of Excellence for Environmental Decisions.

References

Blackmore, L. and G.J. Doole (2013) 'Drivers of landholder participation in tender programs for Australian biodiversity conservation', *Environmental Science and Policy* 33: 143–53.

Boxall, P.C., O. Perger and K. Packman (2012) 'An experimental examination of target-based conservation auctions', paper presented to the 56th Australian Agricultural and Resource Economics Society Conference, Fremantle.

Chapman, A.D. (2009) *Numbers of living species in Australia and the World*, Australia Biodiversity Information Services, Toowoomba.

Connor, J.D., J.R. Ward and B. Bryan (2008) 'Exploring the cost effectiveness of land conservation auctions and payment policies', *Australian Journal of Agricultural and Resource Economics* 52: 303–19.

Department of Sustainability and the Environment (DSE) (2006) *BushTender—the landholder perspective*, Department of Sustainability and Environment, Victoria.

Doole, G.J., L. Blackmore and S. Schilizzi (2014) 'Determinants of cost-effectiveness in tender and offset programs for Australian biodiversity conservation', *Land Use Policy* 36: 23–32.

Fitzsimmons, J. and G. Wescott (2001) 'The role and contribution of private land in Victoria to biodiversity conservation and the protected area system', *Australian Journal of Environmental Management* 8: 142–157.

Golan, A., G.G. Judge and D. Miller (1996) *Maximum entropy econometrics: Robust estimation with limited data*, Wiley, New York.

Latacz-Lohmann, U. and C. van der Hamsvoort (1997) 'Auctioning conservation contracts: A theoretical analysis and an application', *American Journal of Agricultural Economics* 79: 407–18.

Latacz-Lohmann, U. and S. Schilizzi (2005) *Auctions for conservation contracts: a review of the theoretical and empirical literature,* SEERAD, Glasgow.

Marsden Jacob Associates (MJA) (2010) *Review of the Environmental Stewardship Program,* Marsden Jacob Associates, Melbourne.

Moon, K., N. Marshall and C. Cocklin (2012) 'Personal circumstances and social characteristics as determinants of landholder participation in biodiversity conservation programs', *Journal of Environmental Management* 113: 292–300.

Morrison, M., J. Durante, J. Greig and J. Ward (2008) *Encouraging participation in market based instruments and incentive programs,* Land and Water Australia, Canberra.

Morrison, M., J. Durante, J. Greig, J. Ward and E. Oczkowski (2011) 'Segmenting landholders for improving the targeting of natural resource management expenditures', *Journal of Environmental Planning and Management* 55: 17–37.

Pannell, D.J. (2008) 'Public benefits, private benefits, and policy intervention for land-use change for environmental benefits', *Land Economics* 84: 225–40.

Pannell, D.J., G.R. Marshall, N. Barr, et al. (2006) 'Understanding and promoting adoption of conservation practices by rural landholders', *Australian Journal of Experimental Agriculture* 46: 1407–24.

Reeson, A.F., L.C. Rodriguez, S.M. Whitten, et al. (2011) 'Adapting auctions for the provision of ecosystem services at the landscape scale', *Ecological Economics* 70: 1621–7.

Schilizzi, S., and U. Latacz-Lohmann (2007) 'Assessing the performance of conservation auctions: an experimental study', *Land Economics* 83: 497–515.

Schilizzi, S., and U. Latacz-Lohmann (2012) 'Conservation tenders: Linking theory and experiments for policy assessment', *Australian Journal of Agricultural and Resource Economics* 57: 1–23.

Stoneham, G., V. Chaudhri, A. Ha and L. Strappazzon (2003) 'Auctions for conservation contracts: An empirical examination of Victoria's BushTender trial', *Australian Journal of Agricultural and Resource Economics* 47: 477–500.

Whitten, S.M., A. Reeson, J. Windle and J. Rolfe (2007) 'Designing conservation tenders to support landholder participation: A framework and case study assessment', *Ecosystem Services Journal* 6: 82–92.

Whitten, S. M., A. Reeson, J. Windle, and J. Rolfe. (2013) 'Designing conservation tenders to support landholder participation: A framework and case study assessment', *Ecosystem Services* 6:82–92.

Windle, J. and J. Rolfe (2007) *Competitive tenders for conservation contracts*, Central Queensland University, Rockhampton.

Windle, J. and J. Rolfe (2008) 'Exploring the efficiencies of using competitive tenders over fixed price grants to protect biodiversity in Australian rangelands', *Land Use Policy* 25: 388–98.

Zammit, C. (2013) 'Landowners and conservation markets: Social benefits from two Australian government programs', *Land Use Policy* 31: 11–16.

22

Improving the performance of agri-environment programs: Reflections on best practice in design and implementation

David Pannell

Key lessons

- We should be guided by experience. Agri-environment programs have been run over many years in many countries. They provide lessons of success factors (and barriers to success) that should inform how we design new programs and projects.
- Key elements of best practice relate to:
  - the design of programs/institutions;
  - the design of projects/investments;
  - how investment options are ranked;
  - how uncertainty is managed;
  - how people's biases, preconceptions, and self-interest are managed; and
  - how transaction costs are managed.
- Delivering best practice requires expertise. Agencies with responsibility for agri-environment programs should foster the development of expertise in these issues amongst their staff.

Figure 22.1: A mixed agricultural landscape showing linear plantings, patches of remnant native vegetation, and plantings in the middle of the paddock.
Source: Photo by Dean Ansell.

Research and practical experience with agri-environment programs around the world provides many lessons on what leads to success or failure. New programs are often designed without sufficient awareness of these lessons, resulting in lost opportunities to achieve more valuable outcomes.

In this chapter, I outline key elements of what I believe can be identified as best practice in the design and implementation of agri-environment programs. The recommendations are derived from various reviews of programs (e.g. European Court of Auditors 2011; Pannell and Roberts 2010), published guidelines (e.g. OECD 2010) and 15 years' experience working closely with a range of agencies and organisations responsible for management of natural resources and the environment (e.g. Seymour et al. 2008; Roberts and Pannell 2009). I divide these issues into six sections:

1. The design of programs/institutions;
2. The design of projects/investments;
3. Ranking projects/investments;

4. Managing uncertainty;

5. Managing people's biases, preconceptions, and self-interest; and

6. Managing transaction costs.

Depending on the political and administrative context, some elements of best practice may be difficult or impossible to achieve. For example, I note below the importance of long-term funding for many (probably most) environmental projects to deliver their intended benefits. However, given the political and administrative realities of the Australian Government, long-term funding arrangements for environmental projects are exceptionally rare. Perhaps the only example, the Environmental Stewardship Program, was shut down after only four years because it clashed fundamentally with administrative arrangements and culture, and because its virtues were not recognised. (See Chapter 3 for a description and reflection on the Environmental Stewardship Program.)

I have chosen not to exclude recommendations that may be incompatible with some government contexts. However, I have noted those that are likely to face the greatest challenges.

Design of programs/institutions

Additionality: Agri-environmental programs should aim to avoid paying farmers for undertaking actions that they would have done in any case. In other words, managers need to evaluate whether the benefits generated by a program investment are additional.

Continuation after investment ends: Where a program is intended to provide only temporary support to farmers (e.g. in all Australian programs, but not in European programs), it is important to ensure that the actions being supported are attractive enough that farmers will continue to undertake them once funding ends. Otherwise the investment has no enduring benefit.

These first two principles combine to mean that, where support will be temporary, perhaps the only defensible role for agri-environmental payments is to encourage farmers to get experience in a new practice that they are likely to be keen to continue once funding ends. The practice might be something new of which farmers are currently unaware, or which becomes more attractive to farmers with experience.

Appropriate institutional delivery: In some agri-environmental programs, responsibility for overseeing some or all on-ground delivery of projects is devolved to regional organisations. This has been the case in all of Australia's major programs since the late 1990s. In these cases, the program needs to be designed in a way that provides incentives for these regional organisations to respond appropriately. In particular, they should be incentivised to pursue sustained improvements in natural resource outcomes, rather than to support project activities without considering their resulting outcomes. There should be an emphasis on spending program resources well, rather than rapidly. Unfortunately, some of Australia's major programs have generated incentives that directly go against these recommendations. Short time frames for programs and rules that funding will be withdrawn if not spent rapidly enough increase the difficulty of meeting this best practice requirement. (Chapter 5 on environmental NGOs discusses how these organisations can help here.)

Balancing small, moderate and large projects: In programs where the availability of funding is small relative to the amount needed to fund all attractive projects (i.e. in all Australian programs), there is often a temptation to share the available resources amongst a large number of small projects. Sometimes this results in good leverage of program resources, but often it means that almost all projects have inadequate resources and are unable to achieve worthwhile outcomes. This advice sometimes clashes with political preferences to support many projects rather than few. A compromise strategy could be to use a portion of funding (e.g. 25 per cent) to support many small projects to satisfy political needs, and use the remaining 75 per cent to support larger projects that are more likely to be effective.

On the other hand, achieving the most ambitious environmental targets is often disproportionately expensive, with costs increasing dramatically as targets become more ambitious. To maximise outcomes, it may be best to pursue a moderate number of moderate-sized projects, rather than many small or few large projects.

Sufficient time for planning: Program performance is often hampered by a tendency for agencies to delay planning new programs until a previous program has ended or is about the end. Good planning and design of programs and prioritisation of investments requires more time than is usually allowed. Ideally, organisations should commence planning and analysis to develop the next program years before the end of the current program. Even though the scope and parameters of the next program cannot be known in advance, these can be predicted, and sometimes influenced, by the agency to some extent. Having already-analysed investment options ready to put forward can be highly persuasive, and increases the likely environmental benefits generated.

Investment longevity: Finally, funding for agri-environmental programs in Australia tends to be temporary and short-term — typically five years. Environmental problems usually take much longer than this to resolve, so systems for providing long-term funding should be used where possible. If it is not possible to ensure long-term funding, then this should have a strong influence on which projects are selected for funding. In particular, projects that would require significant funding in the long term to maintain the benefits generated by an initial project should not be supported. For example, most projects for control of feral animals or plants would fall into this category because feral animals reinvade once control ends. Similarly, cases where farmers are likely to disadopt practices once funding ends should be excluded. Typically, programs are much too optimistic about ongoing adoption of practices post funding.

Design of projects/investments

SMART targets: A number of agri-environmental programs have been criticised for failing to establish appropriate targets (e.g. European Court of Auditors 2011; ANAO 2008; Park et al. 2013). Specifically, targets should be SMART (Specific, Measurable, Achievable, Relevant, and Time-Bound) in order to facilitate monitoring and evaluation of a program, and to ensure that the funded investments are focused onto suitable activities (see Chapter 4 on setting SMART targets).

Sufficiency: Many projects funded in agri-environmental programs are not designed in a logically consistent way. They are consistent with a project logic, but only in a qualitative sense. They fail when assessed against quantitative questions, such as 'are the funded activities sufficient to achieve the intended land-use changes?', or 'are the intended land-use changes sufficient to achieve the desired natural resource outcomes?' A good project logic is more than a description or diagram of connections between elements of the system being managed or influenced; it quantifies the connections and makes assumptions transparent.

Selecting the right policy tool: There is a tendency for little thought or analysis to be put into the selection of policy mechanisms to be used in a project or program, resulting in inappropriate choices in many cases. In Australia, there is too much reliance on extension in situations where it cannot deliver the desired outcomes. For example, Australia's national salinity program between 2001 and 2007 relied mainly on extension to encourage farmers to change their practices, but the practices being promoted were not attractive to farmers on the required scale and so were adopted to a very limited extent — much too limited to achieve the program's goals (Pannell and Roberts 2010). On the other hand, in Europe and the United States, financial payments are almost the only mechanisms used, often funding activities that are not additional. The framework of Pannell (2008) helps organisations to evaluate the type of mechanism that is most suitable for a particular project. (See also Chapter 18 on the choice of tools depending on public benefits and private benefits arising from an investment.)

Sometimes programs specify which policy mechanisms will be used by projects prior to identification of the projects, and then project investments are selected without considering whether they are suitable for the predetermined policy mechanism. Preferably, policy mechanisms should be selected to match the type of projects that will be necessary to achieve the desired program outcomes. They should be project specific, to some extent. As noted in Chapter 18, Australian programs tend to rely too much on extension and too little on the development of technology.

Ranking projects/investments

Prioritisation: Where funding is limited, prioritisation of investment options is essential. The quality of the prioritisation process can make a major difference to the natural resource outcomes delivered (e.g. Barry et al. 2014).

Rank projects: Programs should prioritise projects, not problems, regions, or issues. Some programs prioritise regions or issues without defining projects, which means that it is not possible to properly consider issues of project cost, project risks, project benefits, or time lags, all of which should be factored into the prioritisation process.

Rank according to value for money: Projects should be ranked according to their value for money — a measure of their benefit divided by their cost (see Chapter 15 on designing cost-effective agri-environment schemes). Failure to do this is one of the most serious mistakes that can be made when ranking projects, but unfortunately it is common. Some systems fail to consider costs entirely, some do include costs but fail to divide by them, and many include only some of the costs that should be considered. For example, it is important to factor in long-term maintenance costs, since they vary so much between different projects, but few Australian systems do so. If maintenance costs are needed but are not expected to be provided, then project benefits should be scaled down accordingly in the ranking process.

Measure the gain against a counterfactual: When ranking projects, benefits should be estimated from the predicted difference in natural-resource outcomes with those without the proposed investment (see Chapter 19 on counterfactuals). Although this seems like common sense, Maron et al. (2013) found that 15 out of 16 systems in use for ranking biodiversity projects failed to do this correctly.

Incorporate all the benefits and risks: There are many factors that could be considered when estimating the benefits of a proposed project. The essentials are the potential values generated, the likely level of adoption/compliance with the project by landholders (Pannell et al. 2006; chapters 12 and 13), various risks that might result in project failure (technical, social, financial, and managerial risks), and time lags until benefits are generated.

Use a sound metric: A commonly neglected issue is how to combine the variables that determine benefits and costs into a metric for ranking projects. Most metrics in use are theoretically unsound and provide poor rankings, even where project information is accurate. Indeed, as is discussed in Chapter 17, the level of benefit delivered is more sensitive to the quality of the metric than the quality of the information fed into the metric. Potential benefits from investment are very sensitive to the use of inferior ranking metrics. Chapter 17 on metrics and Pannell (2015) outline the requirements for a sound ranking metric.

Figure 22.2: INFFER team members discuss the protection of environmental assets with landholders in north central Victoria.
Source: Photo by Geoff Park.

Managing uncertainty

When decisions about project funding are being made, uncertainty about those projects is usually high. Common areas of major uncertainty include the technical feasibility or effectiveness of the proposed actions to be funded by the project, and the valuation of those environmental benefits that are generated. Uncertainty should be accounted for in several ways.

Identify key uncertainties: Project proponents should be required to identify key uncertainties, and to specify what will be done in the project to reduce them.

Carry out feasibility assessment and pilot studies: Projects above a certain scale should be subject to rigorous feasibility assessment before longer-term funding is committed. Funding to support information collection, perhaps in a pilot study, should be provided for six to 12 months, after which longer-term funding should be conditional on the results obtained.

Learn from early experience: Projects and programs should be managed in an adaptive way, with information collected during early stages of the program or project being used to inform changes in management, or even cessation in some cases. In practice, few programs operate with that degree of flexibility, resulting in the continuation of poorly designed investments after their faults are apparent.

Managing people's biases, preconceptions, and self-interest

Acknowledge the values that people bring with them: Being human, the people involved with agri-environment programs are subject to biases, preconceptions, and self-interest (consider the discussion of the different ways restoration and conservation are valued in different places in Chapter 10), all of which can reduce program performance. If managers are aware of these human traits, they can introduce systems to limit their negative impacts.

Optimism versus realism: A pervasive problem is the tendency for people to be overly optimistic about proposed projects. It is common to see proposals in which the benefits are exaggerated, and the costs, risks, and time lags are underestimated. Several factors contribute to this, including vested interests, wishful thinking, and a failure to recognise all relevant difficulties and risks that are likely to affect a project. The ideal strategy to overcome this problem is serious independent expert review of project proposals, but this is only justifiable where projects are sufficiently large. This is another factor that tends to favour moderately large projects over small projects.

Self-blindness: People involved in allocating program funds commonly perceive their existing prioritisation processes as being of high quality. For example, in a survey we found that staff from most regional natural resource management (NRM) organisations believe that their processes are better than average — clearly an impossibility. In reality, the majority of prioritisation processes I have examined have had serious problems. The common belief that they are strong is an impediment to the improvements that are needed. Addressing these misperceptions required strong leadership and participation in appropriate training, and it may be assisted by appropriate signals and incentives built into the program.

Equity versus effectiveness: Proposals to target investment in high-priority projects sometimes meet resistance in the form of arguments that this is inequitable — that resources should be distributed widely amongst many projects on the grounds of fairness. If natural resource or environmental outcomes are desired, these arguments should be resisted, as they can have a serious adverse effect on the achievement of those outcomes. A case built on maximising environmental benefits can readily be built.

Managing transaction costs

When considering potential improvements to the design and implementation of agri-environmental programs, there is a balance to be struck between improving natural resource outcomes and increasing transaction costs (Pannell et al. 2013b; Chapter 16). The most detailed rigorous approaches are only worth the transaction costs involved for relatively large projects. To limit overall transaction costs I have two suggestions.

Beware many small projects: Avoid having the program being dominated by numerous small projects for which an investment in information and analysis cannot be justified. Such programs have little prospect of delivering and demonstrating genuine natural resource benefits.

Start broad, finish deep: When evaluating potential investments, adopt a strategy of starting broad and finishing deep. In the early stages of the process, you can consider numerous potential projects, but each is evaluated in a relatively simple way that requires low transaction costs. This simple procedure is used to eliminate most of the projects from consideration. In the final stages, consider a relatively small number of project proposals, but require them to be developed in a rigorous way to allow sound decision-making about them.

Final comments

Improved natural resource outcomes from agri-environment programs are needed and expected (Audit Office reports have pointed out a repeated failure by the Australian Government to demonstrate environmental outcomes from NRM investment, ANAO 2008). Generating these outcomes is readily achievable with sufficient will, leadership, and attention to the issues raised here.

If you are involved with the design or implementation of an agri-environment scheme or program, can you answer the questions set out in Box 22.1? If you can't, have you considered what this might mean to the success of your project or program?

Box 22.1: Key elements of good design.

A summary checklist of the issues raised in this chapter. Can you answer yes to the following questions?

1. Designing programs
   - Would farmers have adopted the desired practices even without the program?
   - Will farmers continue their adoption of the new practices once program support ends?
   - Are the institutions that are responsible for program delivery incentivised to pursue outcomes?
   - Is the typical project size large enough without being too large?
   - Is there adequate time for planning?
   - Will the practices being promoted require ongoing funding that the program is unable to provide?

2. Designing projects
   - Does it have appropriate targets?
   - Are the project activities sufficient to achieve its targets?
   - Does it use the right policy tool?

3. Ranking projects
   - Are actions (projects, not problems, issues or regions) being ranked?
   - Is ranking based on value for money?
   - Are benefits being measured against a counterfactual?
   - Are all relevant benefits and risks being factored in?
   - Is a robust metric being used for the ranking?

4. Managing uncertainty
   - Have the key uncertainties been identified?
   - Have feasibility assessments been done?
   - Can we learn from the early stages of implementation?

5. Managing people's interests
   - Has independent expert review been undertaken to balance over-optimistic expectations?
   - Have efforts been made to deal with self-blindness?
   - Have arguments for equity undermined the effectiveness of the program?

6. Managing transaction costs
   - Does the program support projects that are too small to justify the transaction costs needed to deliver and demonstrate benefits?
   - Does project selection start broad and finish deep?

Of course, success requires recognition that there is a body of expertise that needs to be mastered, as described above. Agencies with responsibility for agri-environment programs should foster the development of this expertise amongst their staff.

INFFER (the Investment Framework for Environmental Resources) has been designed to streamline the implementation of many of the recommendations presented here (Pannell et al. 2012, 2013a; and see Box 4.1).

The knowledge and experience is there. It is within our power to improve the performance of agri-environment programs.

Acknowledgements

Thanks to the ARC Centre of Excellence for Environmental Decisions and the Australian Government's National Environmental Research Program Environmental Decisions hub for funding support.

References

ANAO (2008) *Regional delivery model for the Natural Heritage Trust and the National Action Plan for Salinity and Water Quality*, Australian National Audit Office Audit Report No. 21. Available at: www.anao. gov.au/Publications/Audit-Reports/2007-2008/Regional-Delivery-Model-for-the-Natural-Heritage-Trust-and-the-National-Action-Plan-for-Salinity-and-Water-Quality.

Barry, L.E., R.T. Yao, D.R. Harrison, U.H. Paragahawewa and D.J. Pannell (2014) 'Enhancing ecosystem services through afforestation: How policy can help', *Land Use Policy* 39: 135–45.

European Court of Auditors (2011) *Is agri-environment support well designed and managed?* Special Report No. 7, European Union, Luxembourg.

Maron, M., J.R. Rhodes and P. Gibbons (2013) 'Calculating the benefit of conservation actions', *Conservation Letters* 6: 359–67.

OECD (2010) *Guidelines for cost-effective agri-environmental policy measures*, OECD, Paris.

Pannell, D.J. (2008) 'Public benefits, private benefits, and policy intervention for land-use change for environmental benefits', *Land Economics* 84(2): 225–40. Available at: dpannell.fnas.uwa.edu.au/ ppf.htm.

Pannell, D.J. (2015) *Ranking environmental projects*, Working Paper 1507, School of Agricultural and Resource Economics, University of Western Australia. Available at: ageconsearch.umn.edu/ handle/204305.

Pannell, D.J., G.R. Marshall, N. Barr, et al. (2006) 'Understanding and promoting adoption of conservation practices by rural landholders', *Australian Journal of Experimental Agriculture* 46(11): 1407–24.

Pannell, D.J. and A.M. Roberts (2010) 'The National Action Plan for Salinity and Water Quality: A retrospective assessment', *Australian Journal of Agricultural and Resource Economics* 54(4): 437–56.

Pannell, D.J., A.M. Roberts, G. Park, et al. (2012) 'Integrated assessment of public investment in land-use change to protect environmental assets in Australia', *Land Use Policy* 29(2): 377–387.

Pannell, D.J., A.M. Roberts, G. Park and J. Alexander (2013a) 'Designing a practical and rigorous framework for comprehensive evaluation and prioritisation of environmental projects', *Wildlife Research* 40(2): 126–33.

Pannell, D.J., A.M. Roberts, G. Park and J. Alexander (2013b) 'Improving environmental decisions: A transaction-costs story', *Ecological Economics* 88: 244–52.

Park, G., A. Roberts, J. Alexander, L. McNamara and D. Pannell (2013) 'The quality of resource condition targets in regional natural resource management in Australia', *Australasian Journal of Environmental Management* 20(4): 285–301.

Roberts, A. and D. Pannell (2009) 'Piloting a systematic framework for public investment in regional natural resource management: Dryland salinity in Australia', *Land Use Policy* 26(4), 1001–10.

Seymour, E., D. Pannell, A. Roberts, S. Marsh and R. Wilkinson (2008) 'Decision-making by regional bodies for natural resource management in Australia: Current processes and capacity gaps', *Australasian Journal of Environmental Management* 15(4), 211–21.

23

Conclusion — Elements of good design

Dean Ansell, Fiona Gibson and David Salt

Breaking News — Australia's national agricultural lobby, Farmers for Farmers, have just signed a historic accord with the conservation lobby, Conservationists at Large, pledging a redoubled effort to renew the natural values of our national farming estate. What makes the accord particularly noteworthy is that the federal government has acknowledged the importance of this new consensus and has pledged $3 billion over five years to reverse the rising rate of extinctions, and declining quality of our land and water resources. The investment will be made primarily through a ramp up of the country's agri-environment schemes. 'This is a once in a lifetime opportunity', says the prime minister.

Of course, this is a hypothetical news story, but you never know what lies around the political corner. The Decade of Landcare announced in 1989 was not anticipated by many in the years preceding it. While it was well received by all and sundry, it did not produce the level of enduring environmental outcomes that was expected (see Chapter 7 by David Salt).

Perhaps that is not surprising. Back then, our understanding of community-based natural resource management (NRM), robust environmental frameworks, market-based instruments, and environmental accounting was basic at best. A quarter of a century later, these fields have developed enormously, and we now have innumerable case studies to reflect on and learn from.

Figure 23.1: A native tree planted in a farm paddock in south east Australia.
Source: Photo by Dean Ansell.

So, when the next big opportunity to conserve biodiversity on farms comes around, will we be able to show that we have learnt from our experiences in agri-environmental policy? What are the key factors a policymaker needs to consider when designing and delivering an agri-environment scheme? Each chapter in this book provides valuable lessons and insights that policymakers should keep in mind when developing agri-environment schemes. We discuss here six central themes that have emerged from the discussions contained in the previous 22 chapters.

Additionality

Agri-environment schemes arguably have two main goals: (1) to shift certain agricultural practices and behaviours towards more environmentally sustainable alternatives; and (2) in doing so to protect or enhance environmental values. Consideration of both is critical. Faced with the decision of where to invest our scarce conservation funds, the decision maker should follow the mantra of any savvy investor and ask the question: 'where can I maximise my returns while minimising the risk?'

Additionality is a term used to define the size of the effect, or the amount of benefit, resulting from an action. In the context of agri-environment schemes, we can think about additionality from two different but equally important perspectives. The first concerns farmers' adoption of on-farm activities.

Consider a farmer who adopts an environmentally desirable agricultural practice and receives payment from an agri-environment scheme as a result. If the farmer had not received a financial incentive, would she have undertaken that specific practice anyway? If the answer is no, we would say that the benefits of the scheme are additional. If the answer is yes, however, we would say that the farmer has received a windfall — a payment for something she was going to do regardless of the scheme.

The extent of additionality achieved by agri-environment schemes varies widely. The USDA Agricultural Resources Management Survey for 2009–2011 showed that additionality in conservation payments ranges between 56 and 88 per cent, depending on the type of scheme (Claassen and Duquette 2014). In other words, for some schemes, close to half of the farmers receiving payments would have undertaken the particular action without the payment. An evaluation of several different agri-environment schemes in France showed that the complexity or scale of change required in farming practice influences additionality. The additionality of more complex measures such as a shift from conventional to organic farming was typically much higher than that of more simple measures (e.g. changing crop diversity) (Chabé-Ferret and Subervie 2013).

We can also think of additionality in terms of environmental outcomes, and ask whether the scheme has led to any change in conservation value or ecological condition. In Chapter 20, Phil Gibbons stresses the importance of focusing on those conservation actions that provide the highest additional benefits, specifically the greatest biodiversity gains relative to the status quo. In doing so, he challenges the traditional focus on investing in the conservation of high-quality habitats on farm land and instead advocates emphasis on 'smaller, more modified remnants that are more vulnerable to loss' and which provide the greatest biodiversity gains as they are starting from a lower ecological condition.

A key challenge lies in identifying and measuring additionality, be it during the planning of an agri-environment scheme and selection of sites, or in retrospect during evaluation of a scheme's effectiveness. Both are important and contribute to improving the efficiency of conservation expenditure. However, as David Duncan and Paul Reich point out in Chapter 19, the consideration of additionality (through a comparison of results with and without the investment) is lacking in the evaluation of Australian agri-environment schemes. They also note that some decision makers hold the false perception that the use of counterfactuals in the evaluation of agri-environment schemes adds considerably to the cost of evaluation. They argue that simplified designs that ignore the counterfactual represent a waste of resources, as their results are unreliable. For an example of cost-effective monitoring and evaluation of agri-environment schemes, the reader is encouraged to review the work of David Lindenmayer and colleagues on the Environmental Stewardship Program (Lindenmayer et al. 2012).

Longevity

Program and project longevity is an important ingredient in designing effective agri-environment schemes. In Australia, agri-environmental schemes tend to be temporary and short-term — typically five years or less. Longevity refers to how long a particular agri-environment scheme (or program) needs to run to be successful. It refers to two different things: (1) whether a scheme is run for long enough to induce a change in landholder behaviour; and (2) whether it is long enough to achieve environmental objectives.

The first four chapters in Part 1 of this book (the agri-environment in the real world) all commented on the long-term nature of environmental action on private land.

'Achievement of these outcomes requires significant, long-term changes in land use and land management, which come at considerable financial and social cost to farmers', observes Geoff Park in Chapter 4.

In Chapter 2, Graham Fifield further supports this, noting that ongoing commitment to a site is important if the landholder is to achieve a good environmental return on the initial investment.

Emma Burns and colleagues (Chapter 3) describe the Environmental Stewardship Program as a policy innovation that delivered this long-term support, providing payments over a 15-year period, but note the challenge of operating such schemes over multiple political and accounting cycles. Indeed, the designed longevity of this program was possibly both its greatest strength and weakness (and a major reason it was discontinued).

But longevity is not just about the completion of on-ground works, numbers of hectares enrolled into a program, or even ecological benefit. It is just as much about changing the behaviour, attitudes, and values of landholders — a program needs to run long enough for this to occur. As David Pannell points out in his chapter on improving the performance of agri-environment schemes (Chapter 22), Australian programs provide only temporary support to farmers. Pannell notes that, where support is temporary, 'it is important to ensure that the actions being supported are attractive enough that farmers will continue to undertake them once funding ends. Otherwise the investment has no enduring benefit.'

When it comes to program longevity, enduring benefit is an important goal against which to judge policy proposals. And if the provision of long-term funding is not possible, then, as Pannell suggests in Chapter 22, a hard truth should be acknowledged about what should be funded: 'projects that would require significant funding in the long term to maintain the benefits generated by an initial project should not be supported.'

Long-term funding is important to creating enduring ecological and social outcomes, but it also contributes to the generation of human capital (skills and knowledge) and social capital (networks, trust and information sharing). Burns and colleagues concluded their review of the Environmental Stewardship Program in Chapter 3 with the observation that 'a valuable outcome that the Commonwealth secured through this program (in addition to the hectares being managed) was the relationships forged with the contracted land managers and developed with the CSIRO and ANU. These relationships should be nurtured to foster further learning and trust'.

Longevity is also an important characteristic influencing the desirability of a scheme to land owners. When it comes to schemes based on tenders, Graeme Doole and Louise Blackmore found in Chapter 21 that 10-year contracts seemed the most desirable. They noted that shorter contracts (five years or less) often fail to achieve desired outcomes, whereas longer contracts (15 years or more) represent a longer commitment than farmers are willing to accept.

Further, farmers may require a higher price to enrol in programs that run for longer time frames and potentially impact on their agribusiness flexibility (Ruto and Garrod 2009).

In Chapter 13, Romy Greiner also commented on how the duration of a scheme influenced the willingness of land owners to participate. She noted that, for the land owners she surveyed, 'graziers were asking for a $0.40 increase in annual per hectare payment ... for an additional year of contract duration'. (This might sound a paltry sum per hectare but keep in mind the properties she surveyed ranged in size from 2,500–10,000 km^2.) This underlines a tension between land owners wanting to participate but not wanting to commit to anything for too long.

Given the short time frames of most programs, the role of environmental non-government organisations (eNGOs) as brokers is critical. As David Freudenberger states in Chapter 5:

> The advantage of engaging a broker is the ability to build lasting relationships to help navigate the complexities and risks of entering and persisting in any market. Many eNGOs have persisted through decades of agri-environment schemes that often don't last for more than one election cycle. Continuity and organisational identity is a strength of many eNGOs.

Policy mechanisms for changing behaviour

The primary aim of an agri-environment scheme is to get landholders to adopt farming practices that deliver improved environmental outcomes. Over the years, a range of mechanisms have been used to try to achieve behavioural change. There have been payments, government regulation (to prevent damaging farming practices), tax advantages, extension (technology transfer, education, communication, demonstrations, support for community network), and development of improved land management options, such as through strategic R&D, participatory R&D with landholders, and provision of infrastructure to support a new management practice. Academic research and on-ground experience shows that the success of these mechanisms in changing behaviour is varied (Pannell et al. 2006). Several chapters in this book point to some of the reasons why.

First, the incentive for participating in a scheme is not always predominately financial. For example, the results of surveys of landholders presented in chapters 12, 13, and 21 reveal:

> [Commercial] farmers rated environmental factors as most frequently influencing their adoption of native vegetation management practices (Chapter 12).

> [F]armers in northern Australia have a high intrinsic stewardship motivation for safeguarding their cattle, land, and biodiversity assets, and that this is fundamentally linked to the pursuit of pastoralism as a chosen lifestyle (Chapter 13).

> Landholders with an altruistic attitude and strong conservation focus, with a relatively low focus on monetary outcomes, are more likely to participate in future [conservation tender] programs (Chapter 21).

It is clear that at least some landholders adopt pro-conservation practices voluntarily, without requiring payments. For example, Saan Ecker in Chapter 12 described a survey of landholder motivations to participate in the Environmental Stewardship Program. She noted: 'Most respondents "strongly agreed" that conservation and enhancement of native vegetation contributed to improved property or landscape health, aesthetics, soil stabilisation, and controlling rising water tables.'

Another example is Chapter 14, in which Maksym Polyakov and David Pannell estimate the extent to which private benefits from native vegetation on farms are built into the price of land, and how those price premiums vary in different circumstances. One potential problem occurring when there are private benefits from conservation is that these benefits are not additional (as discussed earlier).

Another is the problem of 'crowding out', where government funding for a practice reduces the level of unfunded voluntary adoption of that practice by people who are not supported by the program. This can occur if landholders feel that it is unfair for them to receive no recognition for their voluntary efforts while other landholders are receiving payments for the same actions. Graeme Doole and Louise Blackmore in Chapter 21 note that 'tender programs must employ options to counteract crowding out if they are to achieve additional environmental outcomes'. It's not obvious what these other options may be — aside from not to provide incentive payments at all — and is therefore an issue worthy of further research.

We have also seen that flexibility in delivery mechanisms is important. For example, the features of a land management contract may encourage or inhibit landholder participation if certain conditions aren't available. As Greiner states in Chapter 13: 'in general, farmers prefer higher payments, shorter contracts, more flexibility, less accountability and less paperwork.' This is a point supported by Doole and Blackmore in Chapter 21. The importance of each of these features is likely to vary depending on the location, farming system, and characteristics of the landholder. We don't suggest that policymakers pander to these desires — there are public benefits from opposite contract features — rather that they weigh up the public benefits and private costs in delivery mechanisms.

The message here is that agri-environment scheme designers should carefully consider the range of policy mechanisms they use, as some will be more suitable for some groups of farmers than others. Several evaluations of the effectiveness of schemes, both in Australia (e.g. Michael et al. 2014) and around the world (e.g. Gabriel et al. 2010), have found that a one-size-fits-all approach often fails to deliver the best biodiversity outcomes.

Prioritisation

We need to prioritise because there is never enough money available to fund all the available projects. To maximise the environmental benefits delivered by the budget of a program, governments should seek to deliver the best possible value for money. This is done by comparing the benefits and costs of proposed projects and funding those that provide the best return on investment — that is, the highest ratio of benefits to costs (Joseph et al. 2009).

In his chapter on improving the performance of agri-environment programs (Chapter 22), David Pannell provides a checklist of the key aspects of (cost-effective) prioritisation including a focus on projects (actions); ranking according to value for money; using counterfactuals to calculate benefits (as the difference in outcomes with versus without the investment); incorporating all the benefits and risks; and using a sound metric to rank investments. These elements were highlighted separately in several chapters.

Central to the prioritisation process is the explicit consideration of the costs of each project. Failing to acknowledge cost, or failing to appropriately compare costs between projects, has been a major weakness of project prioritisation in the past (Pannell 2013) and is poorly done across environmental evaluation in general (Wortley et al. 2013; Armsworth 2014). This is one of the key reasons that Pannell recommends that prioritisation should be applied to projects or actions, not to different regions, problems, and issues. Only by defining projects is it possible to meaningfully estimate investment costs.

Projects should be ranked according to value for money — a measure of their benefit divided by their cost. In Chapter 15, Dean Ansell points out that the application of this simple principle could result in significant improvements in efficiency in conservation expenditure. He also notes that there is a variety of simple economic tools available to perform such evaluations that remain relatively under-used.

Decision makers should make sure all the benefits and risks are being incorporated. If the level of adoption or likelihood of success is not factored in when projects are being ranked, inferior projects may be selected. Saan Ecker (Chapter 12) and Romy Greiner (Chapter 13) both discuss the importance of understanding the willingness of land managers to participate in agri-environment schemes as being central to the success of the projects included in the schemes.

As Fiona Gibson and David Pannell explain in Chapter 17, the way the metric used to rank projects is calculated and the choice of variables included are important. Errors here can lead to significant losses of environmental benefits. Interestingly, they also show that investing in the collection of accurate information for ranking projects may not be as critical as is often assumed. In many cases, improving the quality of the metric used to rank projects makes a larger impact on the overall level of benefits generated by a program.

Figure 23.2: A failed effort at native revegetation on a farm in NSW.
Source: Photo by David Salt.

Managing risk and uncertainty

As with many types of investment, agri-environment schemes carry significant risks. Chief among these is the risk of failure — primarily the failure to achieve the intended conservation outcomes, which in essence translates to the failure of the scheme. While there are many examples of successful agri-environment schemes, there are many that have failed to achieve their objectives or even led to negative consequences. For example, an evaluation of agri-environment schemes in Victoria found little evidence for benefits to the conservation of reptiles and amphibians (Michael et al. 2014), while a large scheme in Ireland led to an increase in agricultural pests, at the same time failing to achieve its goal of increasing the abundance of the threatened Irish hare (Reid et al. 2007). In Italy, declines in the population of the corn crake, a threatened grassland bird found in farmland, coincided with the introduction of government subsidies for grassland conservation management (Brambilla and Pedrini 2013).

Outcomes such as these may be partly or fully attributed to poor planning and implementation, but are often the result of unexpected ecological response to management. The process of ecological restoration, a primary aim of many agri-environment schemes, is complex and remains poorly understood, particularly in agricultural landscapes where the legacies of past land use and current management and climatic factors create much uncertainty in the response of biodiversity to conservation. This uncertainty not only has the potential to impact on the environmental values delivered from agri-environment schemes, but, as Sayed Iftekhar and colleagues remind us in Chapter 10, also impacts on the adoption of scheme practices. Repeated failures run the risk of alienating farmers and undermining their participation in future schemes.

This underscores the importance of a number of key factors in the design and implementation of agri-environment schemes. In particular, it highlights that identifying specific objectives for conservation is critical in defining and demonstrating success, yet the omission of such objectives is a perennial issue (Hobbs 2007). As Geoff Park outlines in Chapter 4, the use of SMART (Specific, Measurable, Achievable, Relevant, and Time-Bound) targets is crucial in the design of the scheme. Not only do SMART targets play a key role in establishing project budgets and time frames, but they also provide a centrepiece for negotiations with farmers around the aims, feasibility, and risks of proposed interventions.

Several chapters in this book contain ideas and strategies for managing risk in agri-environment schemes. The benefits of starting small as a risk mitigation strategy is highlighted by several authors. We learnt in Chapter 2 that Greening Australia's successful Whole of Paddock Rehabilitation (WOPR) scheme started with a single pilot site, which served not only as a way to assess the feasibility of the approach, but also as a demonstration to farmers interested in the program. As the saying goes, the proof is in the pudding. The Environmental Stewardship Program scheme also started small, focusing on a single target ecosystem and using the outcomes of that initial stage to broaden the coverage of the scheme as it evolved (see Chapter 3).

As David Freudenberger points out in Chapter 5, the level of acceptable risk differs between types of organisations. Non-government organisations, being largely free of the political constraints of government agencies, typically display a higher willingness to fail, and can therefore play a key role in innovation and trialling new approaches.

It is worth remembering, however, that many risks associated with agri-environment schemes, such as the uncertainty in ecological response, cannot be entirely removed. The efforts of the decision maker will be better spent factoring risk into the planning and prioritisation of agri-environment schemes, with various tools available to assist (see the simple metric provided by Fiona Gibson and David Pannell in Chapter 17, which includes the probability of success). The use of an adaptive management framework to identify, respond to, and learn from this uncertainty and unpredictability is strongly advocated by researchers (Lindenmayer et al 2008; Sayer et al. 2013). It should be noted that such an approach brings additional challenges (e.g. funding, expertise), albeit surmountable, for the policymaker.

Above all, understanding, acknowledging, and communicating these risks, particularly the risk of failure, was identified by many of our contributing authors as a critical factor in agri-environmental policy.

Capacity

In Chapter 22, David Pannell provides a list of 22 elements of good agri-environment scheme design. There was a 23rd element put forward by Pannell: 'success requires recognition that there is a body of expertise that needs to be mastered … Agencies with responsibility. for agri-environment programs should foster the development of this expertise amongst their staff.' Which leads us to our final theme — capacity. Capacity is not just about the skills and knowledge contained in the organisations running these schemes; it also relates to the human and social capital found in the regions where agri-environment schemes are being implemented (Curtis and Lefroy 2010).

This book makes it clear that designing, implementing, and managing robust and effective agri-environmental programs requires a range of knowledge and technical skills. For agri-environment schemes to be effective, these skills and knowledge need to be available to policymakers, NRM managers, and the landholders participating in the schemes.

There are a number of ways to develop expertise amongst agency staff. The three approaches recommended by Cook et al. (2013) are: scientists embedded within agencies (internships), formal links between researchers and decision makers, and staff training. Formal links between researchers and scientists are facilitated through various government programs, such as cooperative research centres, Australian Research Council Linkage Projects and programs such as the National Environmental Research Program run by the Australian Department of the Environment. However, as pointed out by Emma Burns and colleagues in Chapter 3, the issue of dealing with scientific knowledge and its application in agri-environment policy within a government department is challenging and will require cultural reform for a more effective integration in future. Attwood and Burns (2012) discuss the disjunct between the spheres of science and NRM policy, suggesting it is systematic in nature. They recommend that scientists need to spend more time understanding the policymakers' bureaucratic and hierarchical system, while the public service structure needs to better reward scientific literacy.

Returning to the issue of landholder capacity, Graham Fifield and David Freudenberger both pointed out in their chapters (chapters 2 and 5 respectively) that landholders and agencies working in the agri-environment need somewhere to turn when things go wrong. Often they seek advice from trusted sources — other landholders, locals, and environmental NGOs they have worked with over time. In recent decades, there have been cutbacks to the level of extension services offered by government (Pannell et al. 2006), and staffing levels of many NRM organisations (Curtis et al. 2014), all of which erodes the capacity of agencies and communities to participate in agri-environment schemes.

In his brief history of agri-environment programs (Chapter 7), David Salt noted that earlier investments in agri-environment programs focused more on building social capital (networks and community groups) and human capital (knowledge and awareness) than targeting specific environmental outcomes. Over time, we have improved our knowledge of what is required to develop programs that will generate these outcomes. It is clear that enhancement of landholder capacity remains an important element of programs, although it is not the only element. There are likely to be benefits from targeting efforts to build capacity to situations where it can make the greatest difference to environmental outcomes.

Box 23.1: The question of value.

The question of value arises throughout this book. It also underpinned much of the discussion at the workshop that gave rise to the book. Whose values are we talking about? Which values do we mean? How do we ensure value for money? Will there ever be enough political pressure for society to adequately protect the multiple values provided by our agricultural landscapes?

During workshop discussions, Rob Fraser, an economist based in the United Kingdom, pointed out that the broader UK society placed a high value on the country's agricultural landscapes. They wanted this landscape to be available for the public to access for recreation, but they also wanted it to be there because it was part of their shared cultural history — even if they never visit it. This led him to raise the issue of different types of value: use values and non-use values.

The use values of an agricultural landscape are the benefits it generates through people making direct use of it, such as for agricultural production (e.g. cropping and livestock activities), or recreation.

Non-use values arise when an agricultural landscape generates benefits even without people making direct use of it. Examples include existence value (the benefit of knowing that the landscape still exists in good environmental condition) or option value (the benefit of retaining the landscape in a condition that does not rule out various options for its future use).

In the UK, much of the agricultural landscape provides a combination of these use and non-use values, with the social-use value of recreation particularly recognised by policymakers. This feature is set to continue into the future, with recent agri-environmental policy changes identifying the need to target areas of land for the provision of recreation values near major urban sites (European Commission 2013).

In Australia, Rob suggested the balance of social values in relation to the agricultural landscape is more towards the non-use value of nature conservation, and less towards the use value of recreation of the UK. It seems likely that non-use values would be considered by many people to be less significant than use values, reducing the prospect of major increases in public funding in the Australian context.

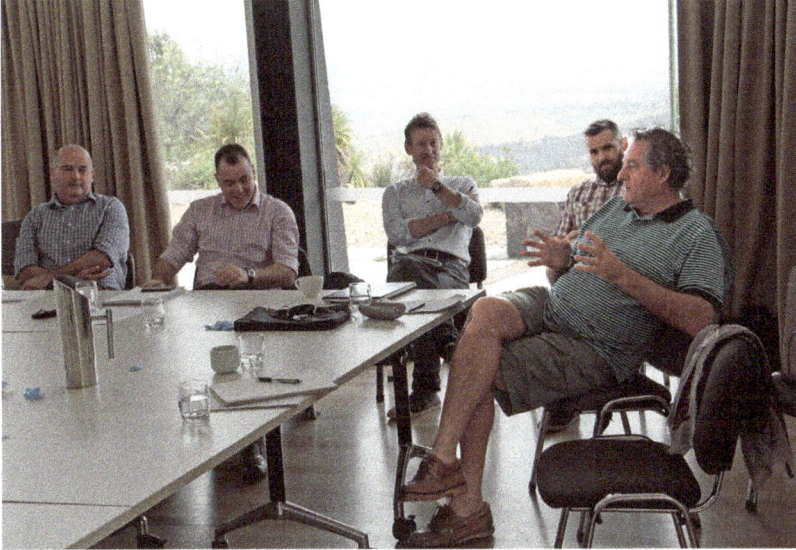

Figure 23.3: Rob Fraser (on the right) at the agri-environment scheme workshop discussing how the UK society values the country's agricultural landscapes. Australia's agricultural landscapes can be glimpsed in the windows in the background.
Source: Photo by David Salt.

Making 'the next big thing' a success story?

Despite the many challenges and criticisms of agri-environment schemes, the fact remains that they represent one of the strongest tools available in the quest to conserve biodiversity in farming landscapes. In our opening chapter, we discussed two contrasting schemes: one focused on restoration (WOPR, see Chapter 2), and the other on conservation (the Environmental Stewardship Program, see Chapter 3), and asked a set of questions about which was better and where the community is most likely to get value for money? The answers to these questions, of course, are 'it depends'.

We would now qualify this context-dependent answer by stating that we believe that the key criteria for successful agri-environmental policy making revolve around our six central themes of additionality, longevity, the application of appropriate policy mechanisms, robust prioritisation, effective risk management, and sufficient levels of capacity.

There are no simple black-and-white answers in addressing these themes, but it is important that the policy designer, implementer, and manager can, at the very least, frame more specific questions against each of them. Our aim in this book has been to help with that framing (and we would emphasise the more detailed list of questions posed by David Pannell in Chapter 22 — see Box 22.1).

If the public mood or political pendulum were to suddenly give rise to a large amount of money being put up for an agri-environment program across Australia, would we as a nation be ready to make the most of it? It is our opinion that we have both the experience and expertise on hand to improve substantially upon past performance.

References

Armsworth, P.R. (2014) 'Inclusion of costs in conservation planning depends on limited datasets and hopeful assumptions', *Annals of the New York Academy of Sciences* 1322(1): 61–76. DOI:10.1111/nyas.12455.

Attwood, S.J. and E. Burns (2012) 'Managing biodiversity in agricultural landscapes: Perspectives from a research-policy interface', *Land use intensification: Effects on agriculture, biodiversity and ecological processes* (eds D.B. Lindenmayer, S. Cunningham and A. Young), CSIRO Publishing, Melbourne, pp. 17–26.

Brambilla, M. and P. Pedrini (2013) 'The introduction of subsidies for grassland conservation in the Italian Alps coincided with population decline in a threatened grassland species, the Corncrake Crex crex', *Bird Study* 60(3): 404–8. DOI:10.1080/00063657.2013.811464.

Chabé-Ferret, S. and J. Subervie (2013) 'How much green for the buck?: Estimating additional and windfall effects of French agro-environmental schemes by DID-matching', *Journal of Environmental Economics and Management* 65(1): 12–27. DOI:10.1016/j.jeem.2012.09.003.

Claassen, R. and E. Duquette (2014) *Additionality in agricultural conservation programs*, USDA Economic Research Service Report No. 170. Available at: www.ers.usda.gov/media/1534156/err170.pdf.

Cook, C.N., M.B. Mascia, M.W. Schwartz, H.P. Possingham and R.A. Fuller (2013) 'Achieving conservation science that bridges the knowledge-action boundary', *Conservation Biology* 27(4): 669–78.

Curtis, A.L. and E.C. Lefroy (2010) 'Beyond threat- and asset-based approaches to natural resource management in Australia', *Australian Journal of Environmental Management* 17: 134–41.

Curtis, A., H. Ross, G.R. Marshall, C. Baldwin, et al. (2014) 'The great experiment with devolved NRM governance: Lessons from community engagement in Australia and New Zealand since the 1980s', *Australasian Journal of Environmental Management* 21(2): 175–99.

European Commission (2013) *Overview of CAP Reform 2014–2020*, Agricultural Perspectives Policy Brief No. 5. Available at: ec.europa.eu/agriculture/policy-perspectives/policy-briefs/05_en.pdf.

Fischer, J., J. Stott, A. Zerger, et al. (2009) *Reversing a tree regeneration crisis in an endangered ecoregion*, PNAS. Available at: www.pnas.org/content/106/25/10386.full.pdf.

Gabriel, D., S.M. Sait, J.A. Hodgson, et al. (2010) 'Scale matters: The impact of organic farming on biodiversity at different spatial scales', *Ecology Letters* 13: 858–69. DOI:10.1111/j.1461-0248.2010.01481.x.

Hobbs, R.J. (2007) 'Setting effective and realistic restoration goals: Key directions for research', *Restoration Ecology* 15: 354–7. DOI:10.1111/j.1526-100X.2007.00225.x.

Joseph, L.N., R.F. Maloney and H.P. Possingham (2009) 'Optimal allocation of resources among threatened species: A project prioritization protocol', *Conservation Biology* 23: 328–38.

Lindenmayer, D., R.J. Hobbs, R. Montague-Drake, et al. (2008) 'A checklist for ecological management of landscapes for conservation', *Ecology Letters* 11(1): 78–91. DOI:10.1111/j.1461-0248.2007.01114.x.

Lindenmayer, D.B., C. Zammit, S.J. Attwood, et al. (2012) 'A novel and cost-effective monitoring approach for outcomes in an Australian biodiversity conservation incentive program', *PloS ONE* 7(12): e50872. DOI:10.1371/journal.pone.0050872.

Michael, D.R., J.T. Wood, M. Crane, et al. (2014) 'How effective are agri-environment schemes for protecting and improving herpetofaunal diversity in Australian endangered woodland ecosystems?', *Journal of Applied Ecology* 51(2): 494–504. DOI:10.1111/1365-2664.12215.

Pannell, D.J. (2008) 'Public benefits, private benefits, and policy mechanism choice for land-use change for environmental benefits', *Land Economics* 84(2): 225–40.

Pannell, D.J. (2013) *Ranking environmental projects*, Working Paper 1312, School of Agricultural and Resource Economics, University of Western Australia. Available at: ageconsearch.umn. edu/handle/156482.

Pannell, D.J., G.R. Marshall, N. Barr, et al. (2006) 'Understanding and promoting adoption of conservation practices by rural landholders', *Australian Journal of Experimental Agriculture* 46: 1407–24.

Polyakov, M., D.J. Pannell, R. Pandit, S. Tapsuwan and G. Park (2015) 'Capitalized amenity value of native vegetation in a multifunctional rural landscape', *American Journal of Agricultural Economics* 97(1): 299–314.

Reid, N., R.A. McDonald and W.I. Montgomery (2007) 'Mammals and agri-environment schemes: Hare haven or pest paradise?', *Journal of Applied Ecology* 44: 1200–8. DOI:10.1111/j.1365-2664. 2007.01336.x.

Ruto, E. and G. Garrod (2009) 'Investigating farmers' preferences for the design of agri-environment schemes: A choice experiment approach', *Journal of Environmental Planning and Management* 52(5): 631–47. DOI:10.1080/09640560902958172.

Sayer, J., T. Sunderland, J. Ghazoul, et al. (2013) 'Ten principles for a landscape approach to reconciling agriculture, conservation, and other competing land uses', *Proceedings of the National Academy of Sciences of the United States of America* 110(21): 8349–56. DOI:10.1073/pnas.1210595110.

Wortley, L., J.M. Hero and M. Howes (2013) 'Evaluating ecological restoration success: A review of the literature', *Restoration Ecology* 21: 537–43. DOI:10.1111/rec.12028.

www.ingramcontent.com/pod-product-compliance
Lightning Source LLC
Chambersburg PA
CBHW050807270326
41926CB00026B/4587